THE SCIENTIFIC PROCESS

THE SCIENTIFIC PROCESS

by

STEPHEN DAVID ROSS

MARTINUS NIJHOFF / THE HAGUE / 1971

ISBN-13: 978-90-247-5026-9   e-ISBN-13: 978-94-010-2987-2
DOI: 10.1007/978-94-010-2987-2

CONTENTS

INTRODUCTION

Some preliminary observations must be made concerning the nature and purpose of this study. What I have attempted here is an essay in the metaphysics of science, and not the "philosophy of science." Rather than concentrating on the details of theory-construction and the formal structure of scientific systems, I have treated science as an *enterprise*, a developing process within human experience. I have used such an approach in order to analyze science in its relationship to other human enterprises, such as art and philosophy, and to clarify its unique goals and characteristics. Often the concepts employed in descriptions of scientific methods are conceived too narrowly; by broadening the focus of attention I have attempted to characterize in a fairly general fashion the goals and methods of science. This has led to formulations which may seem at first glance to depart radically from some "well-established" distinctions of the *philosophy of science*. I hope that it will be clear, however, that such formulations arise at a different level of analysis and concern very different problems from those of the logic of science. In particular, I am concerned with the general goals of science. These must not be confused with the narrower principles of method employed in science at any given time.

The main purpose of this essay, then, is not so much to add to the details of our understanding of science when isolated in its subject matters, goals, and methods from the rest of human experience, as to place the concepts, methods, and aims of science in their relationship to other human activities. Above all, I hope that this essay contributes to an understanding of the sense in which we speak of science as knowledge, and hesitate to do the same for art and philosophy. In this light, it is more an essay in epistemology with special emphasis on science than an examination of the *philosophy of science*. Our culture has become so imbued with science and its fruits that it is essential to take a larger look at the meaning of science, instead of concentrating on its internal and unique characteristics.

THE RISE OF SCIENCE

A. *The Enterprise of Science*

It is useful to begin by considering the subject matter of the philosophy of science; it is not perfectly clear. Apparently we are concerned with *science*, but in what particular sense? Science is an activity, a mode of practice, manipulation, and transformation of material by particular men. It is a body of statements, theories, facts, and laws which somehow gain an existence independent of the material procedures from which they have developed. This body of linguistic utterances has a form which may be abstracted from the concrete contents of the statements themselves. In addition, the activities and experimental techniques of practicing scientists are governed by rules of procedure which also may be abstracted from their concrete acts and techniques. Moreover, there are many and various sciences, with different subject matters and procedures. Is there anything to call "science"? What sort of thing is it? Is any one aspect of it more relevant to what is called "the philosophy of science"?

In general we are well aware of the more obvious characteristics of science. Accounts of scientific procedures, logical conditions, and experimental techniques abound. Many of the conclusions of science have become intrinsic to our cultural environment, both practically and theoretically. It seems foolish to ask "what is science?" Science is simply its methods, rules, criteria, procedures, and conclusions. No other answer is possible, nor should it be sought. Science is a multiplicity of ways of acting, thinking, and asserting. Yet, if we differentiate procedures from conclusions and method from content, where shall we look for connections but to some fundamental quality of the scientific enterprise itself? Is there not a quest for a specific end underlying science? We do, after all, speak of the scientific quest for truth and knowledge.

In a rationalistic age it was possible to conceive science to be motivated by such a quest, for reality and truth were taken seriously as the goals of human reason. But in a more empiricistic time, reality and truth are no longer directly available to us, and become instead consequences of the adoption of methods deemed scientific. We reach truth when our statements are confirmed by *experience*, by evidence obtained through methods of *experimentation* and investigation.[1] Claims to truth are considered legitimate only if they utilize what are commonly accepted as *the* methods and techniques of science. The measure of science becomes its methods alone, and we lose touch with the central quest for comprehension of reality, with the ontological ground that seemed to underlie scientific hopes in other times.

Perhaps we should conceive of science as a total enterprise, a commitment of human energies to the goal of understanding the world in which we live. Such a conception permits us to recognize that this enterprise has its own methods and techniques, that by it we obtain significant results and conclusions which we call "knowledge of the world." But we need not consider science to be its methods alone, its structures, nor even its conclusions, for these are fused inseparably in an enterprise to which its practitioners commit their energies, passions, and hopes. Perhaps with such a conception of science we will not tend to overlook the fact of individual participation in scientific activity out of a preoccupation with the universality of scientific conclusions; nor deny the essential rationality of all scientific theories out of a concern for the grounds of evidence; nor repudiate the active dimensions of science out of a fascination with the logical conditions of the linguistic forms of scientific discourse. Science is much more than a system of statements of a particular form, or a set of rules and criteria applied to certain subject matters. It is a total enterprise, a way of dealing with things. And in fact it is a way of dealing with *everything*, for to any subject matter whatsoever we may bring scientific insight and understanding.

On the other hand, it is not the only way of dealing with the world around us, nor perhaps even the most important. Everything in the world may be responded to aesthetically, manipulated artistically. Whether or not art provides us with *knowledge* of our world, it does

---

[1] "How is the system that represents our world of experience to be distinguished? The answer is that it has been submitted to tests, and has stood up to tests. 'Experience' on this view appears as a distinctive *method* whereby one theoretical system may be distinguished from others." Karl R. Popper: *The Logic of Scientific Discovery*, London, 1939, p. 39. Popper therefore equates "experience" and empirical method. "A theory of the empirical method [is] a theory of what is usually called experience."

give us profound understanding and illumination. Philosophy too stems from unique ways of dealing with things, ways which are not scientific in character. And whether or not religious faith is compatible with the scientific attitude, it utilizes very different methods devoted to very different ends. Not only are these all different ways of experiencing the world, but they are different ways of modifying things: they are distinct, organized modes of human experience. And although they may overlap at times, they involve different modes of functioning and commitment.

The existence of commitment in such enterprises must not be over-looked. As a fundamental mode of human experience, science adopts a total commitment to a world intelligible to scientific understanding. It views the world in its own particular way, selects certain features to be of concern rather than others, chooses certain ways of functioning as its own. Such a way of experiencing may be called a *perspective*,[2] for it involves particular choices and attitudes whether explicitly formulated or not. A scientific act is not simply a particular *kind* of act, but is one with purposes, goals, ends, and conceptions marking it *scientific*. The scientific enterprise is dominated by unique attitudes and selections framing every particular activity as scientific, rather than artistic, practical, philosophical, or theological.

The concept of perspective does not necessarily suggest a relativistic or personal standpoint. Some perspectives are quite shareable and objectively determinable. It does, however, suggest that there are options, there are alternative ways of looking at the world and acting within it. Science is not the sole, nor even the solely defensible perspective. However, if one could maintain that science is the only legitimate procedure for developing a systematic world view, one might hesitate then to speak of options still remaining. The scientific view of things would then be the only legitimate one. Now I do not wish to imply by the *term* "perspective" that this is impossible, nor that such a state of affairs would be nonperspectival. Rather, it would be proper in this case to say that within a more general perspective, science is most valid or worthwhile. Nevertheless, it should be pointed out that we

[2] The term "perspective" is not new to philosophy. I will use it, however, to represent what Justus Buchler refers to when he says that "a perspective is the essence of a judgment [product], the condition and the potentiality of its completion.... A judgment reflects something larger than itself, by virtue of which indeed it is the judgment that it is and has the meaning or communicative effect that it has." *Toward a General Theory of Human Judgment*, New York, 1951, p. 113.

For a more complete analysis of science as a perspective see below Chapter VI, Sections A and B.

have here too simple-minded a view of the validity of science. Despite the testimony of Hume, Kant, and Dewey, there is no primary datum in experience which science alone organizes in rational and legitimate ways. Difficulties arise at the very source of the data.

The conception of primary experience is not quite defensible. It fails to take proper cognizance of the ways in which one's theoretical attitudes affect the collection of the "data." Put more strongly, it ignores the fact that observations are always theory-laden. It is simply not true that all human beings "see" the same world, but interpret what they see differently. In any analyzable sense of "see," we see the world organized and classified in systematic ways. James spoke of primary experience as a "blooming, buzzing confusion": but what is clear about that recognition is that no one really "sees" (consciously experiences) such a confusion, but sees it organized in ways defined by implicit attitudes and principles that may genuinely be called "metaphysical," but which I will simply note are fundamentally theoretical. One observes what has been labelled as *important*, and dismisses from consideration what is irrelevant. Nothing can be seen otherwise. There exists no vocabulary in which to describe "bare" experience – Hume's impressions, or Kant's intuitions. If any of Kant's discoveries must be taken perfectly seriously, it is that experience is always a synthesis of inseparable elements. We seek the simplicity of pure data in experience, but properly speaking we can claim to expose them only at the price of oversimplification – albeit an oversimplification which can have great value.

The colored patch (sense-data) theory of primary experience suffers from a host of well-known difficulties. Paramount among these is the fact that there exists no vocabulary of colored patches in terms of which to describe even visual experience, nor could we even imagine how to do so. Perhaps it is rash to rule out its possibility in advance, but there is little *evidence* to support it, only a mighty wish that things be simple enough to allow it. A further problem, perhaps not so critical, is that sense-data appear to exist only as part of the private and incommunicable domain of personal experience, and do not permit extension to the public world of ordinary objects. The sense-data theory was the major attempt to systematically analyze the world of experience into unquestionably primitive and simple elements. Unfortunately, as Hume himself was forced to recognize in fact (though not explicitly), no adequate knowledge of the world can be based on such simple untheoretical elements. As Whitehead points out, Hume had to introduce the complex organizing principle of *habit* above and beyond sense-data to account for the complexity of human experience. Our perceptual tools are theory-laden to a very great degree. There is nothing that can be called "primary" or simple experience.

Is this so remarkable in ordinary experience? If not, it is then quite obvious in the technical domain of science. When we enter a scientific laboratory, we are confronted by arrays of far from ordinary objects, laid out in far from common ways. What the layman *sees* here is not to be confused with what can be found there by a competent scientist. The scientist *sees* something very different from what is seen by the ordinary man. He selects different factors for consideration.[3] He recognizes what certain arrays *are*. It is not correct to say that he sees what the layman sees, but interprets it in a different way. That does not do justice to how differently men can see the "same" things, depending on their theoretical orientations. One man sees an old woman in a drawing, another a young dancer. Each sees what is there, but they see very differently.

Here I am attempting to elaborate upon the notion of the scientific perspective. But science is by no means the only perspective, and many of the same theoretical determinants can be found in other domains. In music, for example, what is heard depends on one's musical knowledge and sophistication. The trained musician and novice do not "hear" the same thing in any sense – though with an effort of will, the musician might try to hear like the novice. The world is organized by such selections and choices; these are often focused by commitment and dedication. A biologist must *set* himself to view an attractive woman biologically rather than as an object of his affections. Only certain ways of viewing events, by the consideration of evidence, are scientific. The perspective is a wholesale determinant of appropriate data, methods, and goals. A musician responds to music in a trained fashion; he abides by his perspective. So too, science is a perspective, or many perspectives with common attributes.

The claim that science is a perspective or an enterprise rather than a

---

[3] "Ask [the physicist] what he is doing. Will he answer 'I am studying the oscillations of an iron bar which carries a mirror'? No, he will say that he is measuring the electric resistance of the spools. If you are astonished, if you ask him what his words mean ... he will answer that your question requires a long explanation and that you should take a course in electricity." P. Duhem: *La Théorie Physique*, Paris, 1914, p. 218. Also, "the infant and the layman can see: they are not blind. But they cannot see what the physicist sees." Norwood R. Hanson: *Patterns of Discovery*, Cambridge, 1958, p. 17. See also T. S. Kuhn: *The Structure of Scientific Revolutions*, Chicago, 1962 and Stephen Toulmin: *Foresight and Understanding*, Bloomington, 1961.

specific body of statements, or a particular set of techniques and rules, includes the recognition that both the body of statements and the rules governing them undergo modification under the pressures of external events and the commitment to the scientific perspective. The orders of existence that frame human activity are influenced by developments temporally undergone. It is to be expected that any human perspective will be modified by events in history – any other view would be absurdly optimistic and blind to the facts. This is very important. If we maintain that certain techniques and structures are all that is characteristically scientific, then we may claim that the history of their origins and modification is quite unimportant. A historian may be interested in particular ideas of the past, a psychologist may be interested in the origins of knowledge and how it is acquired, but the philosopher need investigate only the character of science, the forms it possesses today. The growth and development of science is irrelevant to the *logic* of science. The only value of history for the philosopher is as a source of examples which reveal the development of logical criteria.[4]

But criteria are often slippery and recalcitrant. Must we forego a consideration of the history of science until we are unerringly sure of what science is? Or can we not learn about science by considering its history? History reveals the way in which modern science has distinguished itself from other modes of human experience, and offers clues to the grounds on which such a separation can be justified. Above all, the rejection of certain problems as irrelevant to science reveals important limitations intrinsic to the perspective of science. It is essential to determine the grounds on which such restriction can be justified.

It must be recognized, for example, that it is too naive to conceive of science as simply the best method for reaching understanding, unless we tautologically mean by this "scientific understanding."[5] Shakes-

[4] The distinction between the history and the logic of science cuts both ways, for it has been pointed out in return: "every student ought to know that the historical and the logical development of science are not the same.... Methodology is a discipline conducted for the most part by logicians unacquainted with the practice of science, and it consists mainly of a set of principles by which accepted conclusions can best be reached by those who already know them." Herbert Dingle: *The Scientific Adventure*, New York, *1953*, pp. 37–8.

[5] Such question-begging is far from rare, especially among those who, because of narrow conceptions of empirical method, wish to rule out all methods other than the scientific as non-cognitive. For example, "a problem would be unsolvable in principle by scientific methods if no conceivable observational evidence submitted to any conceivable logical processing would have any bearing whatsoever upon the correctness of any of its solutions." H. Mehlberg: *The Reach of Science*, Toronto, 1958, p. 247. But there are many other (nonscientific) methods that nevertheless have empirical reference. Mehlberg clearly takes *any* empirical method to be scientific. Does this not beg the entire question of scientific preëminence?

peare, Dostoievski, and Kafka possess a profound understanding of dimensions of human experience, of the possibilities of feeling and response, as well as of literary possibilities and ways of realizing them. Few papers in scientific psychology approach the depth and breadth of their knowledge of man. Freud, whose basic categories are so ill-defined as to be of dubious *scientific* value, nevertheless has wholly remade our conception of man. Poets and painters possess considerable understanding of the skills of their crafts, of the possibilities inherent in words and textures. Philosophy also, though too broad in its conceptions to be scientific, contains profound, if somewhat indeterminate, wisdom of man and the world. These forms of knowledge are unscientific, for varying reasons; but they are ways of understanding things. Science is not the only, nor even the best, way of understanding the world which includes man.

Shall we call science the only *empirical* method of reaching knowledge? How persuasive such a view is, and how misleading! For narrative history looks to the facts as much as the most scientific study, yet without the search for lawfulness essential to science – and lawfulness is more a rational than an empirical quality. And a novelist, however unscientific in temperament, looks to human experience for his subject matter, and must be faithful to it. We cannot define science in terms of empiricism, but must seek out the methods of science and define empiricism in their terms. That is exactly what Hume and Kant in their own ways sought to do.

We must turn, then, to the origins of science, to determine why it is that particular methods have come to represent the single path to knowledge. We wish to determine the conditions which separate science from other great enterprises of human life. Science has been terribly successful in recent years – successful in its own terms as well as in the eyes of the world. The twentieth century is the age of science – having bowed before its supreme authority, and disdained other priests. What are the conditions which underlie its monumental success?

A popular and plausible answer is that science is uniquely successful because it is self-corrective. Yet if we ask for an explanation, we are told that science is self-corrective in submitting to the "facts." It is the great myth of science to be governed by facts. It is a myth because it fails to grasp how the acceptance of scientific hypotheses, whether true or not, conditions the subsequent development of science. It has led, in its most extreme form, to the view that scientific theories are no more than generalizations from past experience, a quite untenable view. It is only in rare and unconsidered matters that past experience is ever summed in a generalization.

There is no denying that science involves a submission to facts, that scientific hypotheses are tested by experimental procedures. Much of the power of scientific methods derives from their reliance on reproducible experimental procedures which by virtue of their reproducibility can be held to be universal and temporally unrestricted. But while we can somewhat reliably *reject* inadequate hypotheses on the basis of such methods, it is far more difficult to justify their *acceptance*. This is the force of Hume's attack on induction. If the claim that scientific method is capable of self-correction is simply a form of scepticism, it merits neither our assent nor our interest. We cannot view the empirical base of science as a *source* of hypotheses, for sheer observation cannot suggest the kinds of hypotheses that have proved successful, nor can it fully warrant them. But if the empirical observations of science serve simply as a test of available hypotheses, it is difficult to see how we ever reach self-correction. We can reject proposals, but what can assure us that we can *correct* them? The empirical structure of science guards us only against too casual acceptance of false principles. It offers no way of reaching satisfactory alternatives.

In "The Fixation of Belief," [6] Charles S. Peirce compares science with other methods of fixing belief, to show the superiority of science. He succeeds, by the rational standards to which he appeals, in showing that the methods of science are the only ones that acknowledge error and seek its elimination. But he cannot prove that in a fundamental biological sense science is best. In fact, the superiority of the methods of tenacity (holding to a belief, come what may) and authority is clear, so far as mere fixation of belief is concerned. By accepting an authority without question, one gains security in belief. If science is superior, it can be so only where unswerving security is abandoned.

If we agree that science is self-corrective – which is not obvious – we must ask why we wish to correct ourselves. Why not remain content with those principles acceptable to authority, or open to pure reason? Why not deny the security of all methods of belief, and take refuge in scepticism? As I will show, it was only after a sense of order and intelligibility had arisen that science properly began. The characteristics

---

[6] Charles S. Peirce: "The Fixation of Belief" in Hartshorne and Weiss: *The Collected Papers of Charles Sanders Peirce*, Cambridge, 1931, 5.358–87. Peirce also recognized that science is unique in that it alone involves a quest for *evidence*. See below, Ch. VI.

of the scientific enterprise are fundamentally historic.[7] The ultimate goals of science are not defensible on some other quasi-scientific grounds, but are rooted in historic developments virtually independent of science itself. The principles of science are normative only if we presuppose essential characteristics of the scientific perspective – the search for and acceptance of evidence.[8]

B. The Beginnings

The Greeks discovered one of the conditions necessary to science: faith in human reason and in the fundamental intelligibility of things. (Even more strikingly, they took the basis of intelligibility to be mathematical.) This faith, it is true, did not then culminate in the systematic and organized body of conclusions and methods we now call "science"; but this is no surprise, for the instruments, the means, and even more important, the conceptions necessary for systematic science had not yet appeared. The Greeks took observations when necessary, and contrived theories to explain what they observed. They relied heavily on empirical evidence and observation.[9] But science as we now know it did not then arise, for science is less a method of confirmation than it is the development of a theoretical organization which systematizes and directs successful investigations. To the Greeks what was unknown was organized according to its importance, not according

[7] Cf. H. Dingle: *op. cit.* "The history of science is inseparable from science itself. Science is essentially a process, stretching through time.... The history of science is science." p. 3.
  Lest it be thought that I am lapsing into historicism here, let me hasten to add that my point is that through its history, scientific investigation has found itself virtually defined by assumptions of such a general character as to be almost without the possibility of defense if challenged — e.g., the faith in th intelligibility of nature, the belief in the possibility of rational explanation. Such general assumptions are usually vague and unclear. Yet without them scientific confirmation is unintelligible. They simply *are* science at some point in history. This is why I claim that they constitute conditions *necessary* to science, though they are susceptible to radical modification under the pressure of evidence. This complexity is the heart of the scientific perspective.

[8] The existentialist assertion of human freedom and outright denial of the significance of scientific methods in ethical and social matters often amount to an explicit rejection of the scientific perspective. Existentialists repudiate the apparent universal appeal of scientific conclusions by simply ignoring them. Science is therefore compelling only if we begin by accepting its perspective. See Ch. VI below.

[9] Anaxagoras and Aristotle, for example, both developed theories of the formation of hail that are models of the consideration of evidence. Insofar as they took into account what was observed they are unmistakably in the scientific tradition. But they may be relegated to a subordinate role in the history of science because their discoveries did not and could not contribute to an organized body of scientific thought. Archimedes too should not be ignored, though similar comments apply.
  Cf. further H. Butterfield: *The Origins of Modern Science*, London, 1957, p. 80. "It was commonly argued, even by the enemies of the Aristotelian system, that that system itself could never have been founded except on the footing of observation and experiment."

to its capacity for successful investigation. They did not recognize the need for particular methods for the solution of "scientific" problems. It was not until problems of the obligations and goals of men, the *causes* of motion, and the underlying principles of the universe proved incapable of universally acceptable solution that science was conceived. Science arose out of an awareness that only *some* problems were capable of determinate solution within a given theoretical system. Galileo explicitly rejected the search for *causes* of motion and settled instead for lawful regularity: only the latter was amenable to his investigations. As he puts it in his *Dialogues*, "it is the purpose of our Author merely to investigate and demonstrate some of the properties of accelerated motion, whatever the causes of those accelerations may be."

The faith in reason prior to the rise of science was relatively undifferentiated with respect to the possibility of attaining success. It was not thought necessary to choose particular subject matters by virtue of their openness to particular methods; energy could be devoted to issues of greater importance in human experience. It was not thought necessary to turn from man to the material world independent of him, for no particular areas of experience were felt to be particularly rational. From this point of view, the central quest of knowledge was for wisdom, for ethical understanding, for knowledge of the ideal forms of political order and of the generic properties of experience and the world, not because these are more truly knowledge, but because they are so fundamental and important. If we have relinquished the quest for knowledge of Being and the Good, and have turned instead to knowledge of brute physical events divorced from human affairs, it is from sheer necessity.

It is possible to merge these very different attempts under the name of "knowledge," for it would not have been at all obvious to a Greek that modern physical science, with all its power and success, is any more worthy of the title than the Aristotelian analysis of Being or the Platonic analysis of Justice. The fact that today most philosophers are inclined to call only the results of science "knowledge" indicates that our conceptions of knowledge have undergone a major transformation, and that we model them now on the structure and methods of science. Of course, one may ask: on what else should we model our conceptions of knowledge? Knowledge is a precarious thing, and only subject matters treated tenderly and scientifically bring forth cognitive fruit. Above all, in what other area of human activity do we reach conclusions that can be held with assurance by everyone?

Yet science too rests on controversy and conflict – that is essential to the proposing and testing of hypotheses. And philosophy, art, and ethics compel *some* degree of assent to their judgments and decisions. Science does not provide certainty nor even wholly universal assurance. And once we realize that some degree of doubt and indecisiveness exists in all forms of knowledge, then we may recognize that philosophy and art also may claim grounds for their judgments and justifications for their claims. Science is not the only human enterprise which offers justification for its conclusions and criteria for its evaluations. The question remains: what is unique about science?

C. The Scientific Attitude

The intellectual period that was dominated by an overpowering, if not total, acceptance of the Aristotelian conception of the universe coincided with a monumental period of religious authority and a sweeping denial of the relevance of certain problems and methods of solution. The Aristotelian cosmology became, with suitable modifications, the accepted view of most authorities in the Europe of the thirteenth and fourteenth centuries. And yet scholastic controversy was encouraged in its pursuit of the finest points of the accepted system. Methods of warfare, the use of projectiles and fortifications, when coupled with the resurgence of Aristotelian cosmology, produced great interest in an adequate theory of motion. The Copernican Revolution was not the profound revolution men have taken it to be because it replaced an earth-centered cosmology by a sun-centered one, but because it rejected "obvious" facts and renewed the possibility of doubting the authoritative account of the world.[10] Giordano Bruno was burned at the stake, not for doubting demonstrable truth, but for doubting authority. The faith in human reason began once more to emerge. Human intelligence did battle once again with authority over who would be master of the human world.

The heliostatic theory became a symbol of the repudiation of authority. That Copernicus rescinded his claim to truth and admitted instead to but rational diversion, a mental exercise, was nevertheless a sign of defeat of the old order, for human intelligence had proved itself capable of fertile imagination, capable of devising alternative

---

[10] The Ptolemaic view was not geocentric but *geostatic*; Copernicus proposed a helio*static* view instead. The key question was, what *moved*? not, what is at the center? All the evidence showed that the earth did not move. Thus Copernicus attacked all available *evidence* to reach theoretical strength.

hypotheses that also "fit the facts." Truth was far less important than this power of the human mind. Copernicus himself never realized the extent to which he had overthrown the older world view. He preserved the notion of crystalline spheres and of a well-ordered, bounded universe. It took men of broader vision to reveal the new possibilities, men like Kepler and Bruno. It took 73 years after its publication for *De Revolutionibus* to be placed on the Index.

The heliostatic theory was clearly not demonstrable, nor was it open to observational confirmation at the time it was formulated. But it was as adequate a hypothesis in terms of evidence then available as was the geostatic theory, and could therefore be used as a weapon against accepted dogma. The devising of alternative theories that are acceptable in terms of available evidence is a significant step in the development of science. For they pose questions that must be answered, and they implicitly attack other, perhaps as yet unquestioned, theoretical assumptions. The great strength of the Copernican world view was that it satisfied the desires of men that the universe be as intelligible as possible, and yet in addition it was capable of constituting a framework for subsequent investigations. It revealed the possibility of doubting what was most obvious (the immovability of the earth) in order to gain theoretical power. Science needed the heliostatic theory in order to further its own development.

Science arose when a distinction was made between problems that within the accepted theoretical framework are resolvable in terms of common observations [11] and problems that do not permit such determinate resolution. Kepler and Galileo did not introduce a new method of knowing; the Greeks and Schoolmen were well aware of the necessity for evaluating hypotheses on the basis of what was observed. The success of Kepler and Galileo stemmed from the fact that they focused on problems that were actually open to resolution at that time, and whose solutions provided theories of great power and breadth. Galileo had first to conceive of the problem of motion in such a manner that it was open to his fertile imagination, by in effect giving up the pursuit of ultimate causes for explanatory power. He turned from the search for a fixed cause (impetus) to the search for an explanation of the variable velocity of a falling body in terms of the motion of repeated attractions dependent on time. He had first to make the conceptual discovery that

---

[11] I mean "common" here literally – that is, observations readily available to anyone who wishes to test the hypothesis in question, for the evidence is reproducible. Science arose when the answer to the demand "show me" became decisive.

velocity was a function of *time*, not distance. In terms of accepted notions this was a revolutionary discovery.[12] With such a conception, the laws of motion were open to his dicovery despite the contrary Aristotelian assumptions and traditional conceptions which arose from the pursuit of far more general matters than a law of motion. It was possible to test hypotheses concerning the motion of bodies by consideration of the facts available, once the notion of functional relationships had been developed, and once enough simplifying hypotheses had been generated. Clearly Galileo could not have considered variations in the force of gravity at different distances from the center of the earth, the resistance of air, nor even the more general problem of bodies with an arbitrary initial velocity. None of these existed in the framework in which he operated. The strength of his approach to the problem of motion was that he isolated it from the encumbrance of traditional problems that were not so resolvable.[13] It is often maintained that Galileo's great insight into scientific method was his appeal to *facts*, his reliance on observation and experiment. Yet facts do not come labelled "facts," and observations must be interpreted. His great discovery was that some hypotheses are testable solely in terms of observations readily available, *once the appropriate theoretical apparatus is taken for granted*.

Consider, for example, the change in theoretical vision necessary for Galileo to see a falling stone as an *event* rather than as a change in state. To Aristotle, the relevant parameters were the beginning and final states and the time elapsed, rather than the velocities themselves and the changes therein. Galileo's discoveries were possible only with a very different sense of motion, as an event to be organized and understood, rather than as a change from one state to another. His very approach to the pendulum was dominated by a Neoplatonic sense of the circularity and symmetry of its motion, rather than by the thought that it is "really" a falling stone approaching rest. This led to his discovery that the period of motion is entirely independent of amplitude, a hypothesis quite impossible to confirm today.[14] Scientific discovery is

[12] Cf. N. R. Hanson: *op. cit.* pp. 37–49. Furthermore, as Hanson shows, Galileo had to overcome the limitations of his geometrical notation.

[13] The abdication of ontology in contemporary philosophy of science reflects the same purposeful attempt to limit scientific problems to such an extent that more complex (metaphysical) matters do not unnecessarily complicate scientific problems. This is not objectionable scientific practice, but it is poor philosophy.

[14] This example I owe to T. S. Kuhn (*op. cit.*) who sets forth a view similar to the one expressed above. Science develops through the acceptance of a general theory whose structure and fundamental problems are relatively well-defined and which is given communal assent. This Kuhn calls a "paradigm." He then points out that "anyone examining a survey of phys-

heavily theory-laden, an outgrowth of antecedently accepted problems and methods of solution.

The appeal to facts is always located in a particular historical and theoretical context. It was fortunate that the new cosmological world view was capable of providing the necessary framework for further scientific investigation. The conception Kepler had of a rational planetary system, and his abiding faith in the rationality of planetary orbits, were what made a formula for the orbit of Mars in terms of Tycho Brahe's observations possible. By contemporary standards Kepler's discovery is far from necessary or even compelling; it is only necessary within the context of its historical presuppositions and theoretical assumptions. In the first place, there is no reason other than historical chance why he need have considered first circles and then conic sections to be rational possibilities. Second, it was sheer good fortune that sufficient observational accuracy was available to permit a choice between an ovoid and an ellipse. Third, there was no antecedent reason to have anticipated that such a problem was more open to successful investigation than ethical and theological problems. It is simply fortunate that all the existing factors combined to produce what could be taken as a solution. Think of how much Kepler had to take for granted (laws of optics, distortions of planetary orbits, coherence of observational data) in order to accept Brahe's data to any reasonable degree of accuracy. Consider the assumptions that must be made about the earth's orbit; and consider that Kepler unquestionably accepted the Copernican conception of a heliocentric and heliostatic solar system. What if the orbit were about some point *near* the sun?

If the orbit of Mars were not a circle, what else *could* it be? In fact, *anything* at all! Kepler's predisposition to classical geometry, his ac-

ical optics before Newton may well conclude that, though the field's practitioners were scientists, the net result of their activity was something less than science. Being able to take no common body of belief for granted, each writer on physical optics felt forced to build his field anew from its foundations." (p. 13) "In the absence of a paradigm or some candidate for paradigm, all of the facts that could possibly pertain to the development of a given science are likely to seem equally relevant." (p. 15) "Only very occasionally, as in the cases of ancient statics, dynamics, and geometrical optics, do facts collected with so little guidance from pre-established theory speak with sufficient clarity to permit the emergence of a first paradigm." (p. 16)

The paradigm creates conditions essential to scientific development. "Normal-scientific research is directed to the articulation of those phenomena and theories that the paradigm already supplies. . . . By focussing attention upon a small range of relatively esoteric problems, the paradigm forces scientists to investigate some part of nature in a detail and depth that would otherwise be unimaginable." (p. 24) "The existence of the paradigm sets the problem to be solved; often the paradigm theory is implicated directly in the design of apparatus able to solve the problem. Without the *Principia*, for example, measurements made with the Atwood machine would have meant nothing at all." (p. 27) ceptance of the intelligibility of the universe, and his preconceptions concerning the nature of that intelligibility were all quite necessary if he was to consider seriously an elliptical orbit. And even then, this conception arose only because the ovoid he was working with could be approximated mathematically only by elliptical approximations. Even a greater degree of accuracy in the data would have prevented his discoveries, for the orbit of Mars is elliptical only to a certain degree of approximation. What is true (according to classical mechanics) is that in an "ideal" setting it would be a perfect ellipse, but how guess this from real data? At another point in history Kepler's conclusions would have been impossible. Yet by the coincidence of factors at that time a law of planetary motion was discovered that virtually permitted no dispute, especially in the light of the later discoveries which it was instrumental in reaching.

Science has come a long way since its inception, but essentially only in its refinement of techniques and methods and in the range and scope of its conclusions. For what seems most clear today is that determinate and specific solutions are available in science upon the development of adequate theoretical organization and preconception. Above all, there is the underlying faith that science will eventually permit a rational decision to be made between any two well-formulated alternative hypotheses by obtaining the appropriate observations or performing the correct tests. If such a decision is impossible, if no way exists to justify a choice between alternative hypotheses, then the issue is considered unscientific. A scientific problem must be capable of determinate solution: it must therefore presuppose sufficient theoretical order to permit such solution. The essential character of the scientific perspective is embodied in its quest for universal assurance, the development of methods to provide every *scientific* problem with a specific and determinate solution.[15]

The sheer acceptance of authority can also produce complete unanimity of belief. Science therefore exists only within a perspective in which authority is forever challenged by a further appeal to facts. This is what suggests that science is forever correcting itself, since it

[15] This is what leads to the repudiation of the search for what Hume calls "the ultimate springs of nature" and to the search instead for general theoretical principles which serve only to unify a scientific system in a comprehensive manner. We cannot really *prove* such principles, so we explicitly repudiate the search for proof in order to preserve the universal validity of accepting them. As N. R. Hanson (*op. cit.* p. 108) puts it: "If you accept the law of gravitation, the laws of Galileo and Kepler, the lunar motions and tides will, as a matter of course, be systematically explained and cast into a universal mechanics.... What could be a better reason?"

constantly reinvestigates the theoretical principles it has found necessary to employ in the past. Every scientific assertion may be tested, however indirectly. Authoritative conclusions, however, are to be accepted
merely because of the sanctity of the testifier. Science begins with a
commitment to the unlimited use of reason and unceasing investigation.

The point to be made clear is that philosophy and art as well as
science also deny the existence of any final authority. They differ,
however, in their ideals of acceptability and validity. In order to pursue any of them we must first accept the challenge of all authoritarian
beliefs. Without a sense of the powers inherent in human intelligence
they are all impossible. But these powers can be conceived in very
different ways, through different goals of determinacy and levels of
generality. Philosophy, art, and science differ in their quest for determinacy and in their range of concern, given the search for methodic
understanding and inventive creation.

D. The Scientific Quest

I have referred a number of times to the *universal assurance* science
seeks.[16] On the face of it, why does science need universality? If one
man discovers scientific truth, is that not sufficient? Can there not be
eternal problems in science which are never resolved? Do scientists not
disagree about the origins of the solar system and the nature of quantum phenomena? How can we be so sure that every question in science
is answerable? Are there not alternative ways of expressing scientific
discoveries which are not actually identical, but which have identical
observable consequences? Finally, if a man denies the scientific perspective entirely, cannot he maintain that he refuses to give his assent to
scientific conclusions? Isn't this another instance of the lack of universality of scientific judgments?

The point is that a hope exists in science which is not realized in its
everyday manifestations. It is true that controversies exist throughout
the sciences, that universal agreement is seldom attained. But science
is the commitment to the pursuit of a method by which all controversies will either be resolved or deemed unscientific, *if the method is accepted*, if we once accept the relevance of evidence to every scientific
dispute. It stems from a faith in human intelligence and formulates a
method and a perspective based on this faith. There are unresolved con-

---

[16] Compare Norman Campbell's definition of science in his short essay *What is Science?*,
New York, 1952, p. 27: "the study of those judgments concerning which universal agreement
can be obtained."

troversies in science, but the expectation is that eventually all such controversies will be resolved or dismissed as irrelevant. Science differs fundamentally in this ideal from art, for example, which pursues multiplicity and diversity. Science rules out of its domain those problems which prove recalcitrant and unresolvable in a determinate manner.[17] It does not matter if universal agreement is never fully obtained in scientific practice; it is a clear and present goal in the scientific undertaking. It is not an unfounded vision, for scientific knowledge does appear to stabilize in some sense; there is a central body of knowledge and experimental techniques that is fully accepted by all who are part of the scientific enterprise. They do agree – they are made to agree by their acceptance of the commitment. (We may further say that evidence compels their agreement.)

Science, then, is not merely a set of statements satisfying certain formal requirements, nor is it a mode of fixed techniques and experimental procedures. It is a complex mode of activity in which statements are made and procedures enacted in order to gain maximal certainty. It is a method of solution of problems which permit solution – that is, in which a determinate choice can be made among alternative solutions. Science pursues the order underlying human experience in terms which entail the ultimate rejection of all alternatives. Science has evolved out of the conception of problems which can be given determinate, if hypothetical solution, and has formulated criteria and methods which are necessary to the recognition and solution of such problems. The scientific perspective marks the development of techniques for reaching determinate solution of those problems which are amenable to such treatment. Such an ideal has always been part of conceptions of reason. Modern science, however, also marks the recognition that some and not all problems permit determinate solution, and pursues these problems exclusively.

It is necessary to stress the resolvability of *scientific* problems because of the strong temptation to suppose that only problems which are capable of precise solution are worthy to be called "problems," open to methodic investigation. The many therapeutic criteria of meaning offered to circumscribe those disputes which are meaningful, reveal the extent to which the scientific perspective has dominated our con-

---

[17] It is worth noting that the special theory of relativity depended heavily on such considerations for its raison d'être. As Einstein himself puts it, " the concept [of simultaneity] does not exist for the physicist until he has the possibility of discovering whether or not it is fulfilled in an actual case." *The Theory of Relativity*, English translation, 1920, p. 22.

ception of rationality. But many human activities are determinate to a considerable degree, methodic and purposeful, and yet are not carried on in the framework of universal compulsion and resolvability. A painter who sets himself the task of creating a significant and valuable painting *cannot* suppose that he is faced with a problem capable of determinate and specific solution. Many possibilities exist, even given the general techniques, tools, and goals he accepts. Artistic activity is methodic and intelligent, but not scientific. Moral practices also, despite our desire for determinacy, have proved themselves open to many valid alternatives. Even philosophic discourse, despite its quest for ultimate satisfaction, seems to function in a context of multiple world-perspectives. Only in science is the variety of alternatives felt to be a serious limitation.

To the extent that diversity is feared, that a single determinate solution to a task presented is conceived to be necessary, the scientific model dominates our conceptions of art, philosophy, and morality. Science marks only one of many modes of human experience, but its success has threatened the others. It is easy to accept a conception of meaning that at best applies to some ideal of science and claim that it represents in general an adequate theory of meaning. The verifiability theory of meaning represents one attempt to formulate one of the essential elements of the perspective of science; in order that problems in science be determinate it is necessary that possible solutions be formulated so that tests of their adequacy be available. Otherwise they may prove to be unresolvable – in other words, unscientific. Such a criterion of meaning is in many ways an insightful representation of the goals and aims of science. That it is not an adequate criterion even for present scientific procedures only emphasizes a point to be taken very seriously – that while the scientific perspective is marked by a quest for universal assurance and ultimate decidability, the existence of diversity and indeterminate alternatives is essential to the *pursuit* of scientific inquiry. There is no reason to suppose that the overriding ideal of science should conform directly to its actual methods. Though science may pursue universal assurance and compulsion, it may operate with dangerously unassured possibilities, *eventually* hoping to make a final choice and resolution. The faith in science is not so much a faith in present knowledge as in what is expected of science in the future. There is a fundamental faith in the future accomplishments of science which characterizes the perspective, and which is not fully captured in the available methods and results.

E. Science as Process

We must, then, distinguish between conditions that are necessary to the *pursuit* of determinate solutions and conditions that govern such solutions. The conditions which govern investigations carried out under a hope of success, and the conditions which govern the faith proved, are not the same. Philosophic analysis tends to take conditions necessary to a proper account of what is known to be necessary conditions of investigations which are still incomplete and open. These are not the same, though they are related by the faith that inquiry will resolve itself. It is not necessary to repudiate this faith to deny that on-going science is the same as finished science, and to recognize that what we are faced with is a continual process, not anything final.

It is possible to call something a "process" and thereby to shirk responsibility for saying anything more about it. In one sense, I am indicating that the characteristics and conditions of science are properties of a process in order to set the boundaries for further discussion. At present I have said little of what this process is but that it contains a central goal, a faith in and quest for problems whose solutions can be made universally compelling and determinate by the collection of evidence. But I emphasize that science is a process for another reason, and that is to emphasize that it is more than a set of fixed rules and techniques, or a body of final conclusions. While philosophers have displayed great insight into many of the necessary conditions of scientific formulation and expression, they have reached very little final analysis. Part of this may be due to human ignorance and limitation, but at least some part derives from the fact that science is not over, that its conditions and standards are not yet fully determined. Science is a process in time in which formulations are tried, hypotheses are proposed, and tests are performed.

The view that science is a process denies that it is possible to permanently separate the conclusions of science and the methods by which they are reached. Every conclusion is fallible, but it must still be used as a premise in further investigation. Nothing in science is finally known, though it must be temporarily taken to be so. There is no definite split between assumptions and conclusions in science; rather, there is a constant movement from one to the other. Science, by means of the manipulation of material and the selection of ideas and objects, works through a process in which conclusions are professed, maintained, modified, and discarded. Nothing is sacred; everything is open to revision. It makes no sense in the consideration of science to emphasize sheer experimentation at the expense of conclusions reached, nor is it plausible to ignore the methods of procedure in the consideration of conclusions alone; for the concluding stage of science is only one stage in the process that is science. The end is not the process itself. We should be particularly impressed by the fact that no conclusions are eternal and final; and if not, the process of their modification is as important as the results of the process. If we recognize that science exists in time and does not attain permanent answers, we must consider in detail both the ontological conditions of science and the means by which science is an active modification of temporary conclusions.

It is, of course, equally necessary to avoid overemphasizing the procedural techniques and changeability of science at the expense of its stable features. Newtonian mechanics has been rejected, but it is still a very good account of certain kinds of macroscopic events. It has been modified, but in some ways very little. It is not very popular in modern philosophy of science to overemphasize the developmental components of science at the expense of the logical structure of the conclusions attained, but it is quite possible to do so.

II

KNOWLEDGE AND FACT

A. The Doctrine of Fact

The classic and rationalist conceptions of the possibilities of reason, that it is capable of grasping or mirroring the nature of things, proved far too optimistic. The world was difficult to know; it resisted simple categorization and analysis. If mind had a native insight into the inner structure of the world, it was forced to undergo far too many indignities of verification and test. Since some of man's most cherished conceptions – the centrality of the earth in the scheme of things, the rational necessity of circular orbits in nature – had proved incorrect, it became difficult to believe anything was either self-evident or capable of assured grasp by the clear light of reason. A statement of knowledge had to be assured and compelling to any qualified thinker if it was to be deemed adequate. Yet the fundamental truths of traditional thought proved quite uncertain – one could quite reasonably conceive a contrary state of affairs. Under the harsh test of "facts," the most sacred truths crumbled and fell. And yet, out of the ruins appeared new truths which no longer permitted blind belief, yet which were somehow more assured.

Rationalist and empiricist conceptions of knowledge are not only alternative conceptions of its nature, but are alternative conceptions of its sources. For a rationalist, knowing is the sheer grasp of truth, once the mind has been sharpened and its ideas clarified. Once ideas have been rendered clear and distinct they can be combined adequately and felicitously. Knowledge is the outcome of a process of active thought in which perception provides at most the material by which ideas are exemplified. The process itself is purely ideational. The laws of logic represent adequate and complete rules for combining ideas in knowing. There is no need to separate knowing and knowledge, for the character of knowledge reflects its sources in human reason.

The empiricist conception of knowing, however, conceives of it as the absorption of diverse and regular bits of experience.[1] The mind enters to combine and associate these bits, and the results of such combination are general expectations of regularity in future experience. Coming to know (which I call "knowing") is not a rational process for Hume, but a natural process of perceiving and associating. Knowledge is a byproduct of living, an accumulation of past perceptual regularities endowing us with expectations of future order. Knowing is the accumulation of discrete impressions; knowledge is nothing more than natural expectation. There is no way of validating knowledge except as it conforms to our expectations and as it represents our perceptual anticipations.

Hume maintains that knowledge can be understood only through its source in experience. But part of his analysis consists of a profound insight into the nature of evidence. For he points out that no evidence can ever logically demonstrate a general conclusion to be true. If we ignore his theory of experience and his conception of expectation, we can simply concern ourselves with the nature of evidence in providing warrant for certain conclusions. If we accept his analysis, no evidence ever conclusively verifies a generalization based on it; no universal law can ever be deemed conclusively verified. But that only faces us with the problem of determining just what relationships do hold between universal and evidential statements, if they support and explain each other.

When we restrict ourselves to an exclusive concern with this kind of question we have restricted ourselves to an exclusive interest in *knowledge* in science rather than with the process of knowing. Science is viewed in only one aspect – that of the conclusions reached and their logical structure. From this point of view, which for reasons that will become clear I will call the *Doctrine of Fact*, the methods of scientific practice, the ways in which hypotheses are proposed and experiments are performed, are simply professional scientific matters, irrelevant to the philosophy of science. Philosophy concerns itself only with the logical structure of scientific systems. In Carnap's early words: "philosophy is to be replaced by the logic of science – that is to say, by the logical analysis of the concepts and sentences of the sciences, for the logic of science is nothing other than the logical syntax of the language of science."[2]

---

[1] By *empiricism* here I mean empiricism in its traditional formulations in Locke and Hume, which are not the most defensible but which are nevertheless the source of the distinction between empiricism and rationalism. It is by no means clear that this distinction is useful today.

[2] Rudolph Carnap: *The Logical Syntax of Language*, New Jersey, 1959, p. xiii.

From this point of view, a scientific system is simply a language with specific syntactical structures and conditions. And not even a language in the broad sense, but a highly restricted, entirely precise and determinate language. Presumably science possessess other properties, but these are rejected as irrelevant to the philosophy of science. For philosophy is but logical syntax and "in the widest sense, logical syntax is the same thing as the construction and manipulation of a calculus."[3] Only those features of scientific statements which are relevant to their function in a calculus are taken to be philosophically significant.

This view in its most extreme form has since been abdicated by Carnap himself, but not its temper or intent. It finds root in widely divergent positions. It leads to the view that scientific explanation and verification are to be understood solely in terms of the formal structure of science. A scientific system is a logical system or calculus in which every statement appears at some level of a deductive order. Lower-level statements are logical consequences of the theoretical statements or hypotheses, and are presumably either directly confirmable by observation or entail statements which are. The sole function of the theoretical apparatus is to permit the deduction of logical consequences. Such consequences are said to be *explained* by theories and laws if the latter, when conjoined with appropriate boundary conditions, entail the consequences. Laws and theories are said to be tested, confirmed, or verified if the consequences which they entail are observed to take place.

The exclusive concern of the philosophy of science with the formulations and linguistic relations of science has shown itself to be deeply insightful into many of the conditions of scientific discourse. It is, however, an intentional restriction to but one dimension of science. It considers the properties and structure of scientific formulations after they have been reached, independent of the conditions under which they arise and are used. It reveals a complete emphasis on knowledge *had*, on conclusions reached, on facts known, rather than on knowing, on the steps preliminary to the adoption of conclusions, and on the function they serve in further investigation. Scientific statements are isolated from their role in the larger process that is science, and studied as to their role in the theoretical system accepted at any particular time. The procedures of science, its social determinants, environmental influences, and personal contributions are set aside out of a concern for the formal relations scientific statements bear to each other.

[3] *Ibid.*, p. 5.

Some important concepts take on very particular meanings here. Notably, *evidence* in the process of science often serves to provide rational grounds for considering a particular hypothesis, and is derived from the material that institutes the problem and the activities that provide possible solutions. But evidence in the logic of science consists of but a set of statements which are logical consequences of a given hypothesis. Evidence in the process of science is derived from experimentation and observation – modes of activity. But when we concern ourselves only with the structures of scientific conclusions, evidence becomes simply those *statements* which sum up the results of scientific activity, isolated from their antecedents, methods, and preliminary procedures. As a total process science includes both meanings of evidence, for there is a movement from evidence which underlies the proposal of a hypothesis, and the methods used to gather new evidence, to the exhibition of the result of such procedures. It is not at all obvious that a particular emphasis on any one phase of science should be pursued at the expense of the others.

Yet logical analysis of science is conceived, not as one phase of a larger process, but as the sole activity with which philosophy should concern itself in its pursuit of the understanding of science. Philosophy, so it is claimed, must not interfere with the creative aspects of science, so it restricts itself to the analysis of the results of the process and their expression. The logical analysis of science is not conceived as an attempt to set forth a theory of the scientific enterprise, but is an account of the structure of the *reports* of science.[4]

B. *A Closed Ontology*

The view that science is no more than a logical system is most felicitous if we imagine a closed universe, a world without change and novelty, which is finished in respects relevant to scientific knowledge. If we knew everything it was possible to know of a closed world we would certainly be able to express our knowledge completely in a determinate and logical system. Likewise, if we were to ask for an explanation of a particular event in a finished and fully understood world we would need to invent no new proposals, nor would we institute methods of

---

[4] "As to the task of the logic of knowledge – in contradistinction to the psychology of knowledge – I shall proceed on the assumption that it consists solely in investigating the methods employed in those systematic tests to which every new idea must be subjected if it is to be seriously entertained." Karl R. Popper: *op. cit.*, p. 31. Contrast this with the words: "when we compare these principles with steps by which the discoveries were actually made, we find scarcely an instance in which there is the slightest resemblance." H. Dingle: *op. cit.*, pp. 38–9.

investigation. Rather, we would give the general rule of which this event was an instance, setting forth in our theory those principles from which everything followed. Science would be nothing but a finished logical system in a closed world.

The logical analysis of science is usually found in contexts which by no means explicitly testify to such closure. Indeed, logical empiricists have even taken the extreme position that scientific statements have no significance other than that of summarizing past experiences. They profess themselves to be totally unconcerned with ontological commitments and presuppositions. They ostensibly make no assumptions concerning the closed character of existence.

If, however, we relinquish the uncomfortable position that science is but a shorthand summary of past experiences, then the formal structure that is a scientific system can be said to be *knowledge* only insofar as it testifies to some features of the world. And the reliance of the logic of science on an unexamined conception of "truth" (or "fact") leads to a closed ontology. The world is thought to be governed by unknown but firm principles: we struggle to reach these principles through science. The structure of a logical system tempts us to suppose that existence is ordered in a similar manner. (At least experience is lawful and closed if existence is not.)

Perhaps the main difficulty is the ancient one that knowledge can only be expressed in terms of determinate and therefore unchanging elements of things. A theoretical system is itself static: it cannot testify to its own development. Science is a passage from theory to theory, not merely the principles which crystallize at any particular moment. If we ignore the developmental character of science, we represent the world by the given theory and the structures of the theory as structures of the world. We cannot understand anything without an implicit theory; we never encounter the world barely and purely. The world *is* for us what our theory-laden interpretations imply. If we conceive of theories as static logical systems, the world will tend to become this for us as well.[5]

The purely syntactical analysis of scientific theories has been rejected by most philosophers of science. Yet there remains throughout a sense of a closed ontology upon which the logic of science rests. For example, an influential view, though sharply criticized in recent

---

[5] "In a logically correct symbolism there will always be a certain fundamental identity of structure between a fact and the symbol for it." Bertrand Russell: "The Philosophy of Logical Atomism," in *Logic and Language*, ed. R. C. Marsh, London, 1956.

times, is that prediction and explanation are formally identical, in that a valid prediction would be a legitimate explanation after the fact and vice versa.[6] The difficulties here, which have been widely discussed, stem from the fact that the particular temporal properties of prediction are usually absent from explanation.[7] The only condition under which prediction and explanation would become formally the same and effectively synonymous is in a universe completed from the standpoint of science. Here a prediction would be no different from an explanation, for the same laws, the same security and lack of doubt would govern both. In a closed universe the tense would not matter, for the future would be closed to investigation.[8]

On the above view, in an incompleted world we can never really "explain" anything, but offer at best *possible* or *likely* explanations; for not only must the explanans *imply* the explanandum, but the statements in the explanans must be *true*.[9] An explanation is always given *after* the fact of knowledge. A likely explanation is no explanation at all.

As I will show, virtually no scientific statements are "true" in the sense taken for granted here. They are not, then, capable of serving as explanans of ordinary events. Consider an "explanation" given for a blowout on a desert highway: the tire was underinflated and became overheated due to rapid driving; by the Boyle-Charles gas laws, the pressure increased proportionately to the temperature creating such enormous pressure that a blow out resulted. (I will take for granted the appropriate statements concerning heat generated by friction and the structural properties of automobile tires.) The key statement in the explanandum is the Boyle-Charles gas law which we know is simply not "true" for real gases. We have, then, no explanation. According to the above conception of scientific explanation, there are virtually no candidates for genuine explanation in science.

A prediction, however, is something quite different. We would not

[6] C. G. Hempel and P. Oppenheim: "Studies in the Logic of Explanation." *Philosophy of Science*, v. 15, 1948, pp. 135-75. Hempel and Oppenheim say only that "it may be said, therefore, that an explanation is not fully adequate unless its explanans, if taken account of in time, could have served as a basis for predicting the phenomenon under consideration," but their position would seem to imply the converse that I have set forth.

[7] Israel Scheffler, who has developed significant analyses of the temporality of explanation and prediction, claims that explanation is temporal in reference like prediction in that one can explain event E only in terms of conditions *prior* to it. Hempel and Oppenheim ignore such temporal restrictions, however, and it is their lack of concern for temporality that I am taking issue with. Cf. I. Scheffler: *The Anatomy of Inquiry*, New York, 1963.

[8] I will argue in Ch. IV that such closure is indeed impossible, implying that prediction and explanation are necessarily (in this world) quite different in function.

[9] Hempel and Oppenheim: *op. cit.*

attempt a prediction in a closed world, nor even in a world in which knowledge was complete and final. Prediction implies not only an openness in time, but an openness of possibility. If the future followed strictly from the present, prediction would be impossible. We would simply state the future as we state the present or past. We do not say that God (or LaPlace's intellect) omnisciently *predicts* the future: rather, he knows it, and can declare the necessary order of things, future or not. Prediction is a genuine concept only in an open world, in a world as yet unfinished, as yet only partly understood. It loses its meaning in a world of complete knowledge. All that is left in such a world is explanation, an account of the structure and order of things.

Many contemporary treatments of time rest upon a similar ontology, for they do not conceive of knowledge as understanding *in* time, but as somehow timeless, outside it. To ask what science tells us about time is simply to ask whether a four-dimensional space-time manifold exists. Such a view denies process entirely, for it treats the world as a completed four-dimensional continuum along which we *by chance* travel:

> The objective world simply *is*, it does not *happen*. Only to the gaze of my consciousness, crawling upward along the life line of my body, does a section of this world come to life as a fleeting image in space which continually changes in time.[10]

In such a world, prediction and explanation are precisely the same, for they express, not a characteristic of a process, but a formal, timeless relation among events.

Consider the function of a calculus in science. "The use of a calculus also ensures that thinking should be completely explicit."[11] In order to render science universally compelling, it must be made routine, the manipulation of straightforward procedures. Once the rules of transformation are mastered, once the calculus is understood, formal procedures are all that is left. Thinking is rendered completely explicit so that it can be validated by others. But when it is *completely* explicit, it can go nowhere: there are no new insights, no new connections to discover, no new properties to evaluate except those which are part of the system.[12] There is a gain in communication and universality, but there

[10] H. Weyl: *Philosophy of Mathematics and Natural Science*, Princeton, 1949, p. 116. I have never been able to decide whether this view implies that it is possible (for us or God) to overcome location in the space-time continuum and thereby to know everything. There is something decidedly Neoplatonic about such a notion. I will pursue this further in Ch. IV.

[11] R. B. Braithwaite: *Scientific Explanation*, Cambridge, 1954, p. 23.

[12] Cf. Ch. III, Sect. E.

is a serious loss in coping with what is completely novel. Nothing outside the accepted system can be dealt with except by a magnificent leap into a new system. The desire to formulate explicit criteria which minimize individual risk presupposes a world sufficiently closed so that risk *can* be minimized and even eliminated.

It is impossible for such a view of science to reflect considerations relevant to knowing. In scientific work proper, especially in new areas, analogies are necessary; concepts must be open; models must be employed. When knowledge is viewed as something possessed rather than something acquired, the only plausible way to exhibit it is as a logical calculus. For now its routine is an advantage; the various concepts and facts known are accessible in a computational system. All knowledge fits together coherently, and can be related economically. But again, only after everything is known.

The view that within a scientific system, statements divide exhaustively and completely into analytic and synthetic, reflects the same preconceptions. The assumption that some statements are immune to empirical test is naive from the standpoint of the actual procedures of science. Statements change their status through the history of science, as the systematic organization of scientific knowledge changes its form. A determinate order can be set forth which fully defines which elements of a theory are open to further investigation and which are not only in a closed system. The growth and development of science renders such an account at best temporary, and at worst a hindrance to further dicovery. The point at which logical analysis of scientific theories becomes *sufficient* (rather than a tool for the process of science) is when the world and science are finished and complete.

The Doctrine of Fact not only conceives of science from the point of view of knowledge reached and held securely, but implicitly extends this to completion. It is one thing to claim that the statements used in the sciences possess formal properties, must conform to certain logical conditions if they are legitimately to be deemed scientific. Presumably, an analysis of these conditions will make an appeal to the actual procedures and techniques of science. The isolation of the analysis of science from the actual procedures of science, however, displays a desire to fix *some* aspects of the world as beyond change. A prediction and an explanation may be conceived as identical only when there is no future into which doubt and insecurity may enter. But the ongoing, insecure – indeed, dangerous – commitments of a creative scientist would then cease to exist. When science is thus routinized, either the ongoing elements are denied, or they are conceived as *unnecessary*. The logical analysis of science disregards what is so important – that science is a way of discovering, attempting, or knowing – and treats only its results, abstracted from its active and methodic procedures. The sense of a world laid out to be known by a secure grasp of unchanging properties dominates, implicit in the view that knowledge can be routinized in a definite set of procedures.

A man building a house can describe its properties, its structure, its geometrical relationships and characteristics. He may also stop working on it when it is "done." But in fact, houses need repair and reconstruction, and may even be entirely rebuilt. Wings can be added; rooms can be converted. The reasons for such modifications are external to the formal structure of the house at any moment. Such a house, in continual reconstruction, always possesses a form. It can always be exhibited structurally by an architect's blueprint. But it is much more than such a blueprint, and its properties are always more than can be expressed by such a formal description. When the blueprint is taken to be a complete account of what the house is, we ignore the process that is the house in time for a cross-section at a particular moment. We adopt a position analogous to the Doctrine of Fact. This has its value, for such a blueprint gives valuable and significant information. It does not, however, permit us to say that we fully understand the house and its development, how its structure has been modified, or the conditions under which such change occurs. If we take knowledge as complete, we will never know why our analyses are always inadequate, lacking something essential, unless we are fortunate enough to live in a world which is finished.

A problem which becomes incapable of solution because of the narrowness of its conception in the Doctrine of Fact is the general problem of induction – the problem of determining the grounds of validity of scientific statements. If the structure of a scientific system is deductive, what grounds exist for inferring the truth of scientific statements from evidence about their logical consequences? Or, if we prefer not to characterize as an inference the confirmation of a hypothesis by evidence, how can we justify accepting as true a general law or theory on the basis of evidence local in time or space? Whatever answer is given must clearly indicate the recognition that such laws and theories *are* valid,[13] in some very important sense.

---

[13] I will use the word "valid" to denote the general sense of having met relevant tests to avoid committing myself to some view of truth. Such validity is not to be confused with the logical validity of arguments (which is one type of validity in the extended sense I am giving it).

Induction was conceived by Hume as a way of knowing, not merely as a method of logical confirmation. We do not wish to affirm a static relationship between known statements, but to be able to affirm and hypothesize new general statements based on evidential material. Hume shows that this cannot be justified in terms of logical relationships alone; no such inferences are permissible. Yet such procedures are essential to the success of scientific methods. If we are not to reject generalizations out of hand as fictions (and on illegitimate grounds of poor ontology), we must analyze knowing rather than knowledge to understand what is involved in induction. Induction was proposed by Mill as a way of making discoveries, not as a way of exhibiting conclusions already obtained. To conceive of induction as but a logical confirmatory relation violates the actual inductive procedures of science, and renders them totally inexplicable in certain important respects.[14]

[14] The attempt to treat the problem of induction as solely a *logical* problem leads to an interesting result given the name of "Hempel's paradox." (Karl Hempel: "Studies in the Logic of Confirmation," *Mind*, vol. 54, 1945, pp. 1–26, 97–121.) If a statement is confirmed by any of its true consequents, then even if we take implication in the strong sense of entailment (a "necessary" relationship exists between antecedent and consequent), then "all ravens are black" is confirmed by "this handkerchief is white," for the former implies strictly and validly "this non-black thing is a non-raven." This is further complicated, as Goodman has shown, by the fact that our evidence also confirms "all ravens are blue," and "there are no ravens." (Nelson Goodman: *Fact, Fiction and Forecast*, Cambridge, 1955, p. 72.)

This is untenable and yet seems to be quite unavoidable if we restrict our analysis to purely logical relations. If we conceive of knowledge as something finished, confirmation can be nothing but a linguistic or logical matter. Hempel's paradox would then hold and would not be paradoxical. We experience the paradoxical nature of confirmation only if we recognize the actual process of science – that is, if we recognize that confirmation is not simply a relationship between known statements, but is part of the scientific process.

Too great an emphasis on the logical principles of science tends to suggest that nothing is taken for granted in science except such principles, and that they are *necessary* to it in a unique sense. I will try to show, however, that problems for investigation exist in science only under the presupposition of logical *and* extra-logical conditions, and that the latter are constantly shifting with the focus of the problem under consideration. Every scientific investigation assumes or takes for granted various principles and facts which make it meaningful. The strongest conditions are logical in the sense that they are necessary to *every* investigation. But each particular investigation also takes for granted many other elements, in the properties and forms of the language used, the variables considered relevant, the factors held worthy of consideration, the "facts" assumed to be true in order that verification be possible.

In abstraction from any concrete problem, the statement "this handkerchief is white" unquestionably appears to be evidence for the truth of the statement "all crows are black," as well as for many of its contraries. This is because evidence is never meaningful in such abstract circumstances. Hypotheses we wish to verify are always conceived in highly restricted, particular ways, to be utilized only in very particular contexts. "All A is B" and "All non-B is non-A" may be *logically* equivalent, but only in a vaguely unbounded sense. As I will show, evidence does not exist in such unbounded contexts, however – at least not in scientific investigations. Absolute logical equivalences are relevant and necessary conditions of evidentiality, but they are not sufficient.

What is necessary in addition is the existence of a particular problem concerning which the statements are made. An ornithologist may wish to know if *this* bird is a crow; the color of a

C. *Truth*

Such considerations appear to indicate that exclusive concern for science as a logical system is not adequate if we wish to comprehend science; that it is not possible to justify induction on logical grounds alone; that science as knowing is justified in much more complex ways. Such considerations, moreover, demonstrate that knowledge can never be conceived as final and complete. But after all, should we not have learned this long ago? Nothing in science is fixed and complete. The point is that lip-service paid to the fallibility of science is not and cannot be sincere if it is accompanied by an exclusive devotion to the logic of science. The implicit premise seems to be that while no scientific law can be maintained with complete assurance to be true, many really are. Science is conceived to be a body of real or putative truths; and truths must, of course, preserve their logical relations and conditions. How can we retain the fallibility of science if scientific systems are but logical systems which take for granted the meaning and usefulness of "truth" and "falsity"?

In general, what passes for truth has no meaning except for the Doctrine of Fact, a perspective in which the world is conceived as complete for reason, a finished product from the point of view of a knower. The generally accepted model of truth or of fact depends on a real world "there" to be known. Truth is taken to refer to structures, orders, or relations which can be grasped and secured in knowledge. If a statement is true, it somehow captures *fact*, it grasps and expresses structure and order, it achieves a correspondence with the world. What the world is is what a true statement says. Truth is important precisely in that it matches reality, and if truth is unchanging and final, so is reality.

handkerchief is quite irrelevant. On the other hand, the color of all the other crows he has seen is quite relevant and significant. Logical conditions, by virtue of their abstractness, are not restricted to any one investigation or class of investigations. Relevance is therefore not a logical property, but a property of the particular problem under consideration. Science has progressed rapidly and efficiently by so conceiving problems as to be open to specific resolution. Absolute problems are set aside. There is nothing intrinsically important about the statement "All crows are black" apart from the specific contexts in which it is utilized. It may serve as a solution to certain problems or it may serve other functions. A man may wish to know if *this* color is black and think "crows are black" in order to obtain a good sample of the color in question. The color of a handkerchief is still not relevant, but neither is the color of any particular crow.

In the actual procedures of science, logically equivalent sentences are not identical or unrestrictedly substitutable for each other, for the concepts utilized (when confirmation is still necessary) are not so clearly marked out as to permit unquestioning transition from one to the other. A given concept may be confirmable in one class of sentences but be disconfirmed in some cases which appear to be logically equivalent – thus revealing the need for conceptual clarification. Elementary particle theory is an example of such a case.

There is no need to consider the various traditional theories of truth in any detail. The correspondence theory of truth clearly imagines that a statement is true if it corresponds appropriately with the world. And it maintains, as do all traditional theories of truth, that a statement once true is eternally true; if necessary we add elliptically omitted references to particular times and places. The correspondence is to what is in the world, to the structure of reality which can be known, which can be finally grasped. To conceive that a statement can barely state *fact*, can be simply *true*, can possess an unalterable, simple property of truth, is to assume a world which can be finally known. The only problem facing us is whether our limited capacities can reach this truth. But it is there, in the world, awaiting intelligent and rational comprehension.

The semantic definition of truth is no better in its assumptions. To claim that the statement "p is true" *means* "p" is to assume that there is a state of affairs "p". There are "atomic facts."[15] We still assume that statements capture, refer to, mean, states of affairs, and when they are properly related to states of affairs, they are true. There is implicit a conception of a world represented by statements, even mirrored by language. When the representation is appropriate, there is truth.[16] The problem for science becomes that of gaining the "right" to make assertions; this is taken to be a very different question. In the same way, the very popular assertive or performatory redundancy theses by no means aid in the solution of this problem.[17] Although the view that nothing is added to a statement "p" when we assert "p is true" has its merits, it ignores the fundamental problem of ascertaining the conditions under which a statement is to be approved or reasserted. There are times when it is inappropriate to assert that "p is true," and appropriate to say instead "I accept p" or "I believe p." Most important of all, when a scientist asks "is this hypothesis *true*?" he seeks those special conditions which justify his assertion of the hypothesis, rather than *proposing* or suggesting it (this is, of course, known as *verification*, or *establishing as true*). As Frege put it, "is it not a great result when the scientist after much hesitation and careful inquiry can finally say 'what

---

[15] Wittgenstein: *Tractatus Logico-Philosophicus*, London, 1961, p. 7.

[16] In evading "atomic facts" by appealing to "protocol statements" considerations of the epistemological status of scientific statements are simply ignored; they are not dealt with adequately. The denial that truth is a property only shifts the epistemological problem to the shoulders of the problem of validating assertions.

[17] F. P. Ramsay: "Facts and Propositions," *Proceedings of the Aristotelian Society*, Suppl. vol. VII, 1927; and P. F. Strawson: "Truth," *Proceedings of the Aristotelian Society*, Suppl. vol. XXIV, 1950.

I supposed is true'?"[18] Redundancy theses dislocate the problem of determining the nature and function of propositions from the province of a statement's truth to that of its assertibility, but they do not thereby solve it. Furthermore, they take for granted the meaning of "assertion" in order to speak of assertive redundancy. It is by no means certain that this is quite a clear notion. Do we not really define an assertion as the kind of thing which is true or false? If the concept of assertion is left unanalyzed, we are left with the same fixed ontology seen above. An assertion says or means *something*. The problem of truth arises when we ask "what?" and "how do we know?"

Ordinarily we take as true statements like: "the sun rises in the east and sets in the west"; "I live a half-mile down the road"; and "the book is on the table." Unquestionably such statements *are* "true" in that they validly represent reality or facts. According to the redundancy theses, by calling them true I am only reasserting them; yet the significant problem of truth arises when we are given a grammatically well-formed sentence and want to know "*is* it true?" or at least, "can I assert it validly?" To what state of affairs am I committing myself when I make an assertion? What are the facts which justify the assertion of p (or of the truth of p)? Despite its well-known inadequacies, the correspondence theory of truth at least attempts to specify just what an assertion has to do with the world in order to be true, or uttered validly.

The above statements, however, can be deemed true only if we take into account their vagueness and generality, if we do not demand precision of them. An assertion is always uttered in the context of conditions which determine what may be expected from it, and whether it fulfills such expectations. Under different conditions a given statement will be more or less satisfactory. Even the statement "the book is on the table" is open to rejection as false if we are too demanding. If the book is wet we may say: "don't worry, it won't stain the table; for it is resting on a plastic tablecloth." A statement can be deemed true only in the context of specific demands which are placed upon it. Usually we so completely take for granted the context of these demands that we find it quite unexceptionable to say "it is true that the book is on the table; go and look; you will find it there." The demands are implicit in the situation in which we utter our remark.

Another way of putting this is by introducing the notion of *validity*:

---

[18] G. Frege: "The Thought: A Logical Inquiry," *Mind*, July 1958. Translated from *Bertrage auf Philosophie des Deutschen Idealismus*, 1918–1919.

a human action (a statement, gesture, even a move in chess) is *valid* if it satisfies the demands of a specific domain of human judgment, and is generally to be taken as valid only for that domain (though it must be emphasized that we seldom spell out the boundaries of such a domain – perhaps advisedly – a lapse which creates many of the knottiest philosophic problems). The statement "the sun rises in the east and sets in the west" is valid in the context of choosing a bedroom facing generally west to avoid morning sun. It is *not* valid in the context of determining the location of a buried treasure precisely due west of a given tree. A given move in chess is valid if it conforms to the rules of the game (occasionally we may demand in addition that it be a strong move, but that is in a slightly different perspective). A given line of poetry is valid by virtue of the function it serves in endowing the poem with significance and effectiveness. Insofar as we demand of statements in ordinary language that they be reliable guides to action or thought, they are valid when they satisfy our implicit or explicit demands (when they satisfy the demands of the perspective in question). If we are concerned with reliability, we may call statements which fail us even once invalid, and perhaps reject them or else attempt to specify the conditions under which they are reliable. But we may also be satisfied with general reliability, thus accepting as true generalizations which do not permit rigorous standards of deduction (the informal fallacy of *accident* arises because of this).

It is not difficult to see that the validity of statements or assertions is what we usually call "truth." Strings of linguistic symbols may function as commands, exhibitions, or even poetic utterances. As such, they may be valid in the appropriate context. When they constitute assertions, however, they are open to empirical test, and may then be deemed true or false.[19] A true statement in ordinary language is one which satisfies the demands placed upon it as a statement – usually (though not always) enabling the prediction and control of future experience and action. "I live a half-mile down the road" is *true* by virtue of the control it gives over methods of transportation, timed estimates for travel, or directions given to others seeking the house. As William James saw, such statements are considered true when they *pay off* in cash value under certain conditions. His error lay simply in viewing truth as one sort of *value* rather than one kind of validity.

[19] Unless *logically* true. Analytic statements are valid in a sense very different from statements empirically true precisely because of the very different context of demands which they are expected to meet. It is not appropriate to collect evidence for their truth or falsity because they are not expected to be confirmable.

This view coincides where necessary with redundancy theses, for in asserting "p is true" I implicitly affirm that p will satisfy the demands I make upon it. Sometimes this is equivalent to no more than reasserting p (believe me, the train *will* be on time); sometimes it is equivalent to a performatory "ditto" (yes, *do* believe me); and in the blind case of asserting "what John said was true," when I am ignorant of his statement, I affirm the reliability of his statement with respect to the demands I expect you to place on it (did he say that the train will be on time? As I said, you may believe him; he knows what he is talking about; you will arrive on time).

It is clear that a "true" statement in ordinary language is true only insofar as quite minimal demands are placed upon it. In scientific contexts, however, we often seem to place unlimited demands on our statements (although I will argue that this is so only in a very special sense). Strict logical rigor demands that we consider them to be indefinitely precise, and utilizable in an indefinite number of different circumstances. A genuine scientific statement must satisfy *every* relevant demand. The vagueness and lack of precision that protect ordinary statements from being falsified are considered faults to be eliminated in scientific investigations; and the price to be paid for the elimination of such faults is that when taken to be perfectly precise, virtually all general scientific statements are actually *false* – that is, they have been disconfirmed in some instances. In the face of the demands of science, even well-known scientific laws, much less complex theories and partly formulated hypotheses, are false in the sense that they fail to be satisfactory under *some* conditions (as, for example, Ohm's law fails to hold under conditions of extremely low temperature). If we say that we seek truth in science, and that scientific laws at present simply are not yet true, then we do not explain the lack of concern that results when a particular law is found to be false. Indeed, what is remarkable is that *false* statements are very often *asserted* in scientific investigations, showing that "p is true" in the ordinary sense and "p" are definitely not equivalent in experimental investigations. In explaining the working of a mercury thermometer, we assert the law of expansion of metals as part of the explanans, *even though it is known to be not quite true*. Such a situation permits us to ignore the small deviations of the law from what is true.

The statement "the acceleration of gravity at sea-level is 32 ft/sec^2" is *false*; and if we render it true by "the acceleration of gravity at sea-level is *approximately* 32 ft/sec^2" it becomes scientifically trivial due to the presence of the altogether vague word "approximately." A "true" statement in the usual sense is one capable of being taken for granted (or asserted) in ordinary circumstances, capable of satisfying ordinary demands. These are not the same as the conditions present in scientific investigations. The strict demands of science are much stronger than the rather vaguely defined expectations of everyday life. It is the inherent vagueness and indeterminacy of ordinary statements that is essential to their being true. When we demand complete precision and determinacy, as in science, we make "truth" an empty word. We must, then, reinterpret the concept of truth in order to satisfy the demands placed upon statements within science.

We might, of course, take the position that truth is properly located in ordinary affairs, that scientific statements are simply not true in any meaningful sense. Yet this would imply that the well-confirmed statements of science are not to be thought of as true *precisely because they are open to indefinite empirical test.* This is a very doubtful position to take. If, when we assert statements to be true (rather than hesitantly uttering the statements themselves), we wish to be sure of them, where else do we find such assurance other than in science? Furthermore, in the important sense of truth or falsity, it is the property of refutability or confirmability that is significant. To deny that scientific statements can be true and yet to still collect evidence to confirm or refute them is to confuse rather than to clarify the issue. It is not sufficient to characterize scientific laws as instruments, approximations, or conveniences. There are too many contexts in which they may actually be *disconfirmed* to justify such claims. The problem is that if scientific laws are held to be true or false in any way, then by strict ordinary standards, they are *false.*

The point is that ordinary standards do not apply in science just as scientific standards are inappropriate in the vaguely defined circumstances of everyday life. It is necessary, then, to define scientific validity for and in science, recognizing that such validity may well differ in important respects from the validity of statements in ordinary affairs. This will be done in Chapter V, Section F.

Anticipating later discussion, I would like to elaborate a little further on the ontological commitments of the Doctine of Fact. Too strict a conception of knowledge presupposes a closed, finished, ordered and regular world. The truth of the world is available; the problem of knowledge is only to represent it and to assure ourselves of it. The only reason experimental techniques are needed is that there is no other way of reaching the world but through our senses and actions. Perhaps in another universe, or under other circumstances, we might be able to directly grasp reality.

I will discuss in the next chapter an alternative ontology, an alternative conception of existence which rests on a method of knowing rather than the existence of "facts." Let me here point out only that when science is conceived simply as a process, as a mode of practice, a human activity, which results in some determinacy, order, control and foreseen regularity in experience, then there need be no world laid out at all, only a world open to human intervention and ordering. There is no need to emphasize the *facts* of the world, nor the immutability of truth. Instead, we may emphasize that knowing is always in a precarious world, in which we try to find order and stability. The "truths" of knowledge correspond here only to the relatively stable features we have grasped amidst the flux. But they are not sacred, for they may be later rejected, after they have endowed thought with order.

Such questions of ontology will have to be put aside until we have examined the Way of Knowing in some detail. For its commitments are very different: its ontological point of view is open and uncommitted to final truth and knowledge. It avoids any presuppositions of completeness. In so doing, of course, it overlooks some important characteristics of the world in which we live, characteristics captured by the Doctrine of Fact. This will now be seen.

III

THE WAY OF KNOWING

A. Activity and Development

The view that science is a complex mode of human activity is not in current vogue. Yet it is rooted in views as old or older than the essentially rationalistic conception of a world whose secrets are open to rational mind. Such views, however, have been either relativistic – Protagoras's claim that "man is the measure of all things" – or sceptical – Hume's denial that human reason is capable of grasping the ultimate springs of nature. So long as an ontology of a fixed and intelligible reality was taken for granted, with knowledge viewed as the grasp or representation of this reality, an adequate analysis of the human role in knowing was impossible. Perhaps as in so many other areas, it was Kant's great insight that man is actively and thoroughly involved in knowing that first opened the possibility of an alternative conception of science. The difficulties with the Kantian view of knowledge are notorious, for he took seriously the possibility of reaching apodeictic certainty in science and yet also maintained the notion of an unknowable world of things-in-themselves underlying experience. Knowledge is restricted to experience; we can know nothing but the course of past and future experience. Only by this restriction can Kant explain the presumed certainty of scientific laws and mathematical theorems. Knowledge ceases to be a rational *grasp* of an external reality and becomes the result of human interaction and modes of thought, a form *given* to the material of experience.

The recognition that the categories of thought are not discovered in nature but are ways of thinking, modes of human functioning, leads directly to the view that knowing cannot be simply a mirroring of the facts of existence. Rather it is the organization of experience, a giving of order to sensory intuitions. A veil forever stands between us and reality and we are restricted to representations of things. Since all we can deal with are phenomena, precategorized and ordered by the necessary laws of thought, we must recognize the essential limitations which bound the closeness we can achieve with nature. We must restrict ourselves to and remain content with the understanding of experience, gained through human participation and activity. The measure of adequate knowing is not its correspondence to a reality outside and independent of human life, but the way in which it orders and regulates experience. There need be no commitment to a fixed reality, for reality is severed from us by the necessary conditions of experiencing and thinking.[1]

The explicit denial of any doctrine of truth and of the existence of a world mirrored or represented in scientific knowledge was, however, of far greater concern to Hegel. Human thought, if it can be said to mirror reality, must be taken as the very model of reality. But human thought never is in fact logical in the classical sense. Rather, it begins narrowly and obscurely and grows out of itself, through its own properties and principles, to increased sophistication and comprehension. The structure of this development is the Hegelian dialectic, the advance of knowing through necessary stages in the advance of thought. Every judgment is always imperfect, extreme in one sense or another. Yet because it is extreme it initiates its own antithesis, creating its own alternative. And though each of these positions is imperfect and insufficient, each is a part of what adequate understanding must include. Thought moves on, in its own momentum, to the next step, a synthesis of these less-perfect views. This new level, however, is but another stage in thought, another partial insight, also imperfect. The process repeats itself forever.[2] Thought does not approach reality or fact, for there is no fact or Truth. The advance of thought through dialectic is the only truth there is. We never actually reach knowledge. We do advance from lower to higher levels of knowing.

The Hegelian view, of course, is developmental and process-oriented rather than static and cross-sectional. It explicitly avoids a commitment to a fixed reality of any sort. There are no eternal governing principles of the world other than the openness of existence and the development of self-consciousness and awareness. There is no reality but the reality which becomes, no truth but that which has come to be from earlier truth, just as there is no knowing but that which develops

---

[1] In fact Kant denies the existence of a fixed reality and places regularity and order solely in experience, for freedom arises as a condition of human existence in a realm of unobservable and therefore indeterminate things. Truth ceases to be a property of statements in relation to underlying reality and refers to an ordering or structural property of judgments in experience.

[2] Not really forever for Hegel, for he foresaw an end to the process. Nevertheless, this is irrelevant to our interests and purposes.

from earlier, incomplete understanding. The world is throughout still becoming. Thought cannot asymptotically approach the facts of existence or truth. It can only grow out of itself. The physical sciences, with their claims to truth, in fact attain only temporary assurance. They take the present state of development to be eternal, and attempt to capture a developing process with a static logic. The very act of predication in classical logic asserts the existence of an unchanging relation between subject and predicate. Such relations simply do not hold in the universe in which we live.

Developmental theories that leave no room at all for the relatively permanent insights of the physical sciences cannot expect to be of great influence in modern philosophy. Some basis surely exists for claiming that facts discovered in science are independent of the men involved in their discovery, however involved men are in the formulation of such facts. Hegel's conception of a logic of development ignores the intimate relationship between science and its commitment to independent natural laws as objects of knowledge. The notion that knowing could grow out of itself was profoundly important, but left no way of accounting for the ever-increasing success of the physical sciences. It was up to followers of Hegel – Marx and Mannheim – to analyze the ways in which knowing is essentially a human process, subject to its inherent commitments and properties, rather than a grasp of an *external* and fixed reality – for they both took very seriously the distinction between knowledge that can be dispassionately viewed because it is remote from human needs and purposes, and social or normative knowledge that is tightly bound up with human values and emotional needs. Physical knowledge is objective and universal *only* because it is value-free and remote. The model of knowledge as a grasp of a reality independent of a knower is not valid, but it is relatively adequate when applied to the physical sciences remote from the basic commitments of the men involved in its pursuit. When, however, we consider economic, political, or psychological judgments we come face to face with the preconceptions, antecedent commitments, and needs of those involved in the cognitive situation. Here it is simply inappropriate to speak of knowledge of fact or reality, for there is no external reality or fact. There is no objective grasp of things in language. Thought does not match fact. It but grows out of its acquired characteristics under the pressure of needs and circumstances. There is no fixed truth about society, only a development from what men once conceived it to be to what they will take it to be in the future, under different conditions. We may pretend that knowledge is independent of the means utilized in attaining it, or at least that truth is so independent. But when we consider truths that matter, from which we cannot remain detached emotionally, we find that the means of knowing intrude irremovably. The "facts" of social and ethical existence are but the assertions we are willing to accept. The model of knowledge mirroring "facts" ceases to apply when we are too close to the essential commitments of the knower.

The conception of science as a process of knowing is not a property of idealistic conceptions of science alone. Since for the empiricist we ultimately rely in knowing on experience, we may well doubt whether we can ever justify the appropriation of truth as some fundamental property of things. We can do no more than assert the internal development of science as it increases our comprehension of order and regularity in experience. The theories we accept at any given time are chosen because they sum up and function significantly in a process of knowing (in the process of *coming* to know), not because they are factual representations of some reality. Perhaps we should repudiate the conception of fact altogether and emphasize instead the active element of ordered experience in science.

It is worthwhile here to consider in detail one development of the view that science is fundamentally a human activity to discover both its strengths and weaknesses. For however important are the insights of this point of view, it suffers from fundamental limitations. On the other hand, it seems to me that these limitations have only occasionally been clearly formulated. I shall endeavor to remedy this by considering John Dewey's theory of knowledge, to display both its insights and failings. Like the Doctrine of Fact, Dewey's conception of science is basically inadequate, for it too ignores vital features of the scientific process. Reformulation will not rectify the omissions; only a philosophic point of view which avoids the narrow commitments of both of them can do this.

B. Knowing

The Hegelian influences in Dewey's philosophy are manifold and profound. Here it is necessary only to emphasize that Dewey's epistemology is based fundamentally on the concept of *knowing* rather than on knowledge possessed. In fact, knowledge has no meaning divorced from knowing. There is no realm of truth which knowledge can grasp or represent. "Knowledge" is a concept abstracted from the concrete operations and successes of particular investigations. "The conception of knowledge as such can only be a generalization of the properties discovered to belong to conclusions which are outcomes of inquiry. Knowledge, as an abstract term, is a name for the product of competent inquiries."[3] Like Hegel, Dewey is struck by the inadequacy of philosophies which rely on any sense of a fixed realm of truth to which knowledge conforms. Hegel felt that thought develops out of its own inner necessity, that the logical conditions of knowing represent the internal process of the development of thought. Dewey, not an idealist, sees that thought depends on the way men experience their world, and is subject to conditions of existence other than its own development.

Science is essentially a problem-solving activity generalized into a coherent method isolated from discrete problems and demands. It arises in doubt; something is wrong, the situation is seen to need correction. Methods of active investigation are introduced until they produce a successful resolution, a new situation free from the doubtfulness and insecurity. Such a process is inquiry, which results in warranted assertibility. "Inquiry is the controlled or directed transformation of an indeterminate situation into one that is so determinate in its constituent distinctions and relations as to convert the elements of the original situation into a unified whole."[4]

The Doctrine of Fact treats science in isolation from its own procedures; it abstracts the results (laws and logical forms) of science completely from its practices. The way of knowing, however, places its emphasis on the things done, the acts performed, in the scientific process. In order to know we must perform acts, manipulate things. Only by instituting change can we achieve knowledge. "The method of physical inquiry is to introduce some change in order to see what other change ensues."[5]

Another way of emphasizing the active nature of knowing and experience is by recognizing that knowing as an activity in the world always changes it to some extent. The Doctrine of Fact conceives of the world as containing features or aspects (laws, forms, regularities) that are static and eternal properties of things. These are *facts* which knowledge attempts to mirror or represent. But when knowing is emphasized in science, it must be viewed as a human activity, one which produces specific changes in the world. The success of knowing is to be measured by

[3] John Dewey: *Logic, the Theory of Inquiry*, New York, 1938, p. 8.
[4] *Logic.*, pp. 103–4.
[5] *The Quest for Certainty*, New York, 1929, p. 84.

the effects it produces. "Knowledge or science, as a work of art, like any other work of art, confers upon things traits and potentialities which did not *previously* belong to them."[6] We can only speak of knowing as an art if we emphasize that it is, like all arts, a mode of practice, with techniques and criteria developed in its own procedures. It is an art of doing, with the intent of making and achieving something. It is not a mirroring of reality; it is rather a process devoted to changing things in certain respects. "The chief consideration . . . lies in the perfecting of methods of action. . . . Regulation of conditions upon which results depend is possible only by doing, yet only by doing which has intelligent direction . . . which plans and executes in the light of knowledge."[7] Knowing is the active transformation of experience.

C. *An Open Ontology*

The conception of science as an activity differs from the view of the Doctrine of Fact in an even more important way than its emphasis on the active elements in knowing. It commits itself to a world differing markedly from a world conceived to possess an underlying realm of truth open to intelligent comprehension. A world in which *knowing* is possible contains discordant and disordered elements as well as satisfactions toward which knowing can be directed. "The union of the hazardous and the stable, of the complete and recurrent, is the condition of all experienced satisfaction as truly as of our predicaments and problems. While it is the source of ignorance, error and failure of expectation, it is the source of delights which fulfillments bring. . . . A purely stable world permits of no illusions, but neither is it clothed with ideals."[8]

Perhaps it is obvious that the world in which we live is a mixture of settlements and problems, of determinateness and confusion. Surely the Doctrine of Fact does not deny the flux and chaos of ordinary existence. It only affirms the underlying intelligibility of things, the possibility of reaching facts amidst the chaos. Dewey's claim, however, is that the world is only sufficiently stable to permit the solution of problems which arise in the confusion. The success of science does not testify to a permanent realm of facts. The value and success of science depend only on the existence of both stability and confusion, insecurity and satisfaction in experience, on but sufficient coherence and regulari-

---

[6] *Experience and Nature*, 2nd ed., Chicago, 1929, p. 311.
[7] *The Quest for Certainty*, p, 36.
[8] *Experience and Nature*, p. 62.

ty to permit the solution of problems which arise. Not only do the sciences not assume that any of their results are final, but they *cannot* do so. As Dewey says, "the settlement of a particular situation by a particular inquiry is no guarantee that *that* settled conclusion will always remain settled."[9] In the sciences, every inquiry is particular to some extent, even though it adds to the whole of our knowledge of the world. And as particular, it cannot give rise to the presumption that its results are final. "There is no such thing as a final settlement."[10]

What is far more important than the conception of the possibilities and limitations of knowing, however, is the ontology presupposed. For not only is knowing a process, but so is experience, and so is the universe. The world is an affair of affairs, a process of processes. And "because we live in a world in process, the future, although continuous with the past, is not its bare repetition."[11] Perhaps Dewey never relinquishes his Hegelian tendency to emphasize the evolution of the universe in time. Perhaps he is attempting to affirm the *necessity* of novelty in the world, out of its own development. But it is more reasonable to interpret the above statement as an affirmation of the openness of reality, its openness not only to change and novelty, but to limitations and ignorance. The world is known and regulated because in the past certain things have been done by men, certain statements have been made, certain events have been intentionally directed. Such activities do not testify to anything fixed in the world, only to the success of certain procedures. The world is open to Dewey in the strongest sense – that novelty is the one assured fact of existence. The world never finishes, never becomes fixed in any of its aspects, but changes throughout.

Such an ontology becomes much clearer if we recognize that knowing is viewed as a *human* activity, that the focus is on human functioning in the world. The Doctrine of Fact presupposes a world which contains an external realm of fact to be known. Such a world exists regardless of human participation and striving. Indeed, the Doctrine of Fact isolates what is to be known from the personal participation of men in knowing, and claims that what is to be known is objective and universal in a sense that isolates it from individual human endeavors. The view that science is a human activity, however, conceives of knowing as but one of many human activities. The world is so structured that

⁹ *Logic.*, p. 8.
¹⁰ *Ibid.*, p. 35.
¹¹ *Ibid.*, p. 40.

men can know it, but this only means that certain kinds of activities are successful in obtaining the solution to certain kinds of problems.

When we consider the world as it might be without human participation and intervention, we pretend to ourselves that such a world is governed by immutable laws, that a realm of fact and truth underlies the flux of particular events. The future is still open, is still indefinite. The physical world is a process, and its future, while it depends on the past, is as yet undone. The physical world is as yet unfixed and unfinished. But to conceive of science as possible is to suppose that the world is intelligible, and we take this to mean that the future is closed from the point of view of truth, fact, or the discoveries of science. The past and present, when known, constitute the future. This is what intelligibility is taken to mean. Knowledge of the world presupposes a set of laws to be known.

The temptations of this position are manifold; we imagine ourselves stepping outside the world to reach its underlying nature. But when we emphasize that knowing is but one of many different modes of experiencing this world, one way of interacting with, appropriating, and transforming the various things which we encounter, it makes no sense to speak of grasping and comprehending this fixed realm of truth. The world becomes open, unfinished, and knowing becomes one way of guiding its future in a foreseen direction. We do certain things, produce certain effects, act upon various elements of our world, and direct it in some rather than other directions. Knowing is a mode of intentional transformation of existence; we develop systems of assertions to facilitate the transformation of existence in foreseen and desirable directions. We act upon the things of our world, make statements about them, perform experiments upon them, and remake them into new elements of a different world. We do not step back from the events of experience in order to know what underlies them, but actively participate in their transformation, and thereby know them. The openness of existence is partly due to the openness of human experience.

In other words, where the Doctrine of Fact commits itself to an ontology of fact underlying and governing the flux of events, the Way of Knowing affirms the openness of the world and the corrigibility of every knowing. The emphasis, however, on corrigibility is stronger than the claim that the rules of evidence prevent us from ever conclusively verifying an empirical statement. For Dewey, knowing never succeeds in grasping truth, for there is no truth to grasp. There are only guideposts to use in every new activity for the directed transformation of experience. Knowing, when viewed as an active modification of things, is always a *particular* act, to be redone under new circumstances. In effect, science ceases to be a body of true statements, and becomes a tool for application in particular situations. To know something is not to utter a true statement about it, but to relate a set of activities, statements, and predictions to a body of evidence and to use it accordingly. Such a complex mode of activity can never be finished. The future is always new.

Traditional empiricism has certainly affirmed the uncertainty of knowledge and the corrigibility of all assertions. But this has always been in the Doctrine of Fact: the facts do exist, but we have no way in our limited experience to reach them with certainty. We can never fully confirm any statement, due to the limitations of experience. We *suppose* that the world is lawful, though we can never be *sure*. Dewey's position, however, emphasizes that the world not only *may* change, but that it actually does. It is a process, and so is knowing. We rely on past evidence for future knowings, for we have no other choice. Knowing directs itself from past discoveries to future control and prediction, but never by discovering ultimate or permanent facts.

The Doctrine of Fact assumes that there is such a thing as a pure and perfect (eternal) fact. But nothing in science is ever permanent; facts are simply *accepted* within an inquiry in order to reach a successful conclusion. They may be rejected in other contexts. For example, we often accept perceptual evidence in physical experiments. We could not inquire at all if we did not. Yet on the other hand, we can also investigate perception itself, as in psychological experiments. There are no sacred and eternal facts here, but conditions we accept in a given investigation in order to specify the boundaries of a problem. Any given "fact" may be itself questioned in another investigation. A given problem is *constituted* by the existence of conditions, by what is accepted. There are no permanent facts, but facts selected for a particular investigation. "Facts . . . are not self-sufficient and complete in themselves. They are selected and described . . . for a purpose."[12]

It seems peculiar to claim that no facts exist. Like the truth of a statement, a fact seems to be a simple property of things. It does not change, nor does it depend on circumstances. But to claim that facts exist in this sense is to affirm a universe *closed in that respect*. Dewey's ontology is open and changing. Statements may be accept*ed* and warrant*ed* (to emphasize the activity involved); but they are not simple re-

[12] *Ibid.*, p. 113.

flections of reality. They are not true by virtue of what they stand for and represent, but by what they do and what can be done with them. The statement "the earth is round" is valid in some contexts, not in others which demand greater precision, or which are concerned with different kinds of issues. Science is always open, always has further to go. Only if it actually settled all problems of existence could we speak of final and unchanging facts, and this would only be possible in a fixed and final world. In the actual operations of science, we take things for granted, accept certain statements as true. These are facts – *of the case*. But only because no "case" could exist unless we accepted certain conditions which define it. This is why knowing is relative to particular situations – it is relative to the conditions taken for granted. If mass had not been conceived as a fixed property of matter, Newtonian physics would have been impossible. It was so conceived, and physical science was born; but this conception was nevertheless an error.

D. *Difficulties*

The Way of Knowing arises out of a reaction to the Doctrine of Fact which makes unwarranted ontological assumptions and which ignores the active components in science. The philosophy of science for the Doctrine of Fact is but a logic of confirmation, in the firm conviction that the rest of science is irrelevant for philosophy. The Way of Knowing attempts to restore omissions and to exhibit a richer and more adequate ontology. However, in practice it tends to become as extreme as the logic of science, and seems to deny many of the most important insights of contemporary philosophy of science. Dewey, in particular, reveals serious problems in his conception of science which should not be overlooked.

The conception of an underlying reality of truth or fact which science attempts to discover may be illegitimate, but it stems from the realization that science does have a goal that is at once more general and more specific in scope than the solution of particular difficulties. Problems do arise in scientific practice, and the goal of scientific inquiry is to solve such problems. But they do not arise haphazardly: they arise out of a pursuit of knowledge of the world, the search for wisdom and understanding. Science is not simply the application of particular statements in ordinary experience – it is devoted to the quest for complete *knowledge of reality*. The Doctrine of Fact at least recognizes the goal of scientific inquiry, and offers this as the essence of the scientific quest.

Dewey, however, does not accept an ontology within which he can speak of facts to be permanently known, nor of truth grasped and held securely. There is no fixed goal of science – only the development of means for the solution of problems of existence as they arise. Such a conception appears on its face to violate the working goals of practicing scientists. But what is far worse is that there seems to be no measure of the success of inquiry. We cannot compare it to reality directly. There are no facts there to be taken and used as evidence. There are only particular and successful modifications of situations to greater satisfaction and dependability.

This leads Dewey to the position that the success of any inquiry depends on a *felt* quality of unification or satisfaction in experience. The problem of knowledge since Kant's insights into the role of man in knowing, has been that of determining how and on what grounds science can grasp even temporary truths – how men become capable of securing knowledge of the world in their limited experience. Given a particular situation, how are we to draw from it universal regularities and principles that are applicable to all of existence? How can we extend our knowledge legitimately into space and time as we must if it can be called "knowledge"? Traditional empiricists assert that there are facts to be grasped through observation. Dewey, however, does not conceive of observation as a simple event. It is rather a complex reflection of human experience, the result of trying to see in this or that way, including the results of previous inquiries. The fundamental warrant for knowing is a *felt* unification of disconnected elements of experience. Each situation confronted by an individual is unique, to be set in order by unique methods. Inquiry is the ordering, the unification, the transformation of a situation. A situation is felt to be doubtful, and inquiry is the institution of methods to remove this feeling.

The difficulty here is that not every irresolution in a situation is a problem for inquiry. Sometimes problems may be enjoyed, sometimes ignored. Investigation arises when problems are responded to as problems – as something to be improved by investigation (a circular situation to say the least). All situations have loose ends – they are doubtful or problematic only when we both have methods for inquiring into them, and when we are set to treat them by such methods. A dial reading to an engineer may institute inquiry – which terminates when the dial reading has changed. It may mean nothing to an ordinary man, or it may form a pleasant pattern to a man of visual sensitivity. Science may develop techniques for further scientific inquiry, but situations become scientific not by virtue of feelings, but by virtue of the techniques that are applied to them. Perhaps feelings exist in all situations, but they are only part of what is characteristic of them. What is primary are ways men *come* to the situation, ways which are governed by the situation itself. A man searching for oil in the desert (engaged in this inquiry) may also be awed and inspired by it. Feelings do matter: but inquiry exists here only because the man is *set* to search and inquire. He operates in a scientific frame of reference by doing certain things. Feelings only contribute part of the structure of such frames of reference.

The point here is that Dewey's emphasis on knowing tends to emphasize the *particularity* of knowing in each particular situation. To a great extent this is legitimate. But science is not merely the transformation of a particular situation. It contains a body of rules, techniques, criteria, *attitudes*, which are applied to a particular subject matter in a particular situation. We are taught to be scientific by learning when and how to apply particular methods and techniques. *Any* situation may be dealt with scientifically, and no situation *must* be treated thus. The transformation inquiry effects in situations is not a transformation of feeling-tone, but an improvement according to distinct criteria developed as part of the procedures of science. This appears to be a vicious circle, but it cannot be broken by appeal to pervasive qualities of situations. Science does involve the pursuit of the comprehension of reality, and develops techniques and leading principles for this. It also involves the transformation of concrete situations by time-tested methods and procedures. Logical relationships are part of science as are feelings of satisfaction at adequate completion of an investigation. Dewey's emphasis on pervasive qualities in situations does not do justice to the ways in which the criteria developed in science determine the validity of proposed solutions. Though we should also keep in mind that the Doctrine of Fact unwarrantedly takes such criteria to be necessary to and permanent in science, Dewey's view simply does not do justice to the necessity of such principles to science as it exists at any time.

The Way of Knowing, when taken in an extreme form, also tends to overemphasize the active, practical results of science. "Inquiry effects *existential* transformation and reconstruction of the material with which it deals."[13] Investigation does transform things – attitudes, comprehensions, meanings. A thing known is different from one not yet

13 *Ibid.*, p. 159.

understood. But the changes science produces are not always *physi-cal*, though Dewey often suggests they are. This is due to an overemphasis on the active components of science, ignoring the theoretical structures. It arises from construing activity narrowly, which Dewey is only guilty of sporadically. He so emphasizes the transformations in science that he fails to pay due attention to the permanence of conditions that are developed. Science is viewed as incomplete without application; it is complete only when entered into concrete physical transformation of human experience. "What is sometimes termed 'applied' science may then be more truly science than is what is conventionally called pure science."[14] A moral value of Dewey's intrudes here which is an essential part of his attitude toward science. Man's dilemma is basically a practical-moral one: how to improve his world and make it better. Science is an activity devoted to this pursuit, and is incomplete when only theoretical.

Here lies the basic limitation of Dewey's conception of science as a process. He seeks a natural function that science can serve in the life of man which will provide its ultimate appraisal and evaluation. Science becomes the transformation of particular situations in human experience. But when science is reduced to instruments available in particular situations, without recognition that the instruments do possess characteristics and principles of their own, it ceases to be itself. We move blindly from unique situation to unique situation, for we have nothing but the results of past investigations to guide us. The sense that science produces something, becomes a body of theories and laws intimately related in a logically coherent form, is set aside. The results and effects of the process of science, the knowledge of the world it achieves, are simply dismissed from explicit consideration.

Although we may agree that all scientific conclusions are tentative and may well be rejected in the future, there is a sense in which legitimate and valid conclusions describe the world in a more significant and profound way than do ordinary superstitions and common-sense aphorisms. The activity of science produces a body of conclusions which are profoundly insightful descriptions of the world in which we live, though we may and probably will eventually reject them. Analogously, if logical principles are considered only to be discoveries of science, it is difficult to see how they can become necessary conditions of scientific investigation. Any method which denies that a scientific hypothesis must be empirically testable is considered unscientific on logical grounds.

[14] *Experience and Nature*, p. 161.

But such grounds exist only if we agree that logical conditions hold and that they do constitute defining conditions of scientific practice. The active conception of science developed by Dewey does not leave room for logical analysis; the *necessity* of logical conditions is not taken seriously enough.

There is no need to further elaborate the particular difficulties remaining in Dewey's philosophy of science. His recognition of the active elements in scientific inquiry is profound and illuminating, whatever its problems. Surely it must be possible to develop a philosophy of science which in some way affirms both the active and formal qualities of scientific investigation. We need not commit ourselves to an ontology of fixed truths when we recognize the relative permanence of scientific laws and methods, nor need we deny the existence of such principles when we emphasize that science is the active pursuit of knowledge rather than a formal body of assertions related in predetermined ways.

E. Other Aspects of Knowing

An additional element must be taken into account in an adequate philosophy of science that affirms the active components of knowing. Dewey, in emphasizing the activity in knowing, does emphasize the human elements, the feelings and preconceptions men bring to science. But he does not deal in detail with the *individual* quality of every investigation. It is possible, however, to lay primary emphasis on the fact that a human *individual* always is present in a scientific investigation; the community of proof in science is not of indefinite extent. In a discovery a man always takes a risk. He asserts his discovery to be more valid than judgments heretofore taken for granted. The Doctrine of Fact with the logic of science attempts to set up general rules applicable blindly in all instances so as to minimize individuality of risk and interpretation. But we cannot legislate the investigator out of existence. If we offer permanent and closed rules for scientific method we close off investigation with them. Every discovery of major importance involves a major revolution in the methods and techniques of scientific practice as well. A scientist who breaks new ground forsakes the accepted community of routine that has already been charted.

Such a view, though not without serious difficulties, is set forth by Michael Polanyi.[15] Polanyi repudiates the attempt to routinize scien-

---

[15] M. Polanyi: *Personal Knowledge*, London, 1958.

tific method, the attempt to predigest all scientific discovery. A discovery in science is a break into new areas out of a conviction that
there can be found truth. A scientist *appraises* evidence in a new way,
as meaning something novel. There is no routine way of making a significant discovery. "The avowed purpose of the exact sciences is to
establish complete intellectual control upon experience in terms of precise rules which can be formally set out and empirically tested. Could
that ideal be fully achieved ... we ... would be relieved of any occasion for executing our personal judgment. We should only have to follow the rules faithfully."[16] This is the Doctrine of Fact, and it clearly
distorts the actual procedures of science. For knowledge is always an
individual act. It is always an individual who accepts or rejects a given
standard. "The selection and testing of scientific hypotheses are personal acts, but like other acts they are subject to rules." [17] A scientist is
a performer; he acts according to certain maxims, but his activities, if
successful, include his personal appraisals.

The reduction of philosophy of science to linguistic matters completely overlooks the individuality of each event, the personality of each
investigator. Like Dewey, Polanyi claims that occasions are unique,
never simply "fitted in" to maxims. "Since every occasion on which a
word is used is in some degree different from every previous occasion,
we should expect that the meaning of a word will be modified on every
such occasion."[18] Linguistic analysis is a guide to inquiry but not its
totality. The individual makes a choice of meanings and faces the world
individually or not at all. "No solution of a problem can be accredited
as a discovery if it is achieved by a procedure following definite rules....
Any strictly formulated procedure could also be excluded as a means of
achieving discovery."[19]

Discovery in science exists in two senses. Contemporary philosophy
of science, with its emphasis on linguistic analysis and specified methods,
in effect claims that all discovery consists in the routine application of specifiable methods. The Doctrine of Fact suggests not only the
existence of definite facts to be known, but definite methods by which
to know them. Modern science is dominated by the image of technical
facility, of routine method, as found in research firms and scientific
organizations. Discovery in this sense is of a fact in the same form as we
have to come to expect from previous inquiries. It is in accord with

16 *Ibid.*, p. 18.
17 *Ibid.*, p. 30.
18 *Ibid.*, p. 110.
19 *Ibid.*, p. 123.

prevalent theories, postulated conditions, and accepted stipulations. It is a fact easily understood and digested.

But as Polanyi points out, scientific progress rests much more upon major discoveries that demand a complete reworking of our conceptual framework. Newtonian physics, the theory of relativity, and quantum mechanics all attacked previously accepted beliefs in fundamental ways. When accepted, they revealed cherished beliefs to be but superstitions. The Doctrine of Fact would reduce science to digested rules thereby *preventing* major discovery from ever taking place. Einstein discovered new facts about the world by risking a new method with new rules, by departing from the conceptual framework of existing theories. As Polanyi states, "traditions are transmitted to us from the past, but they are our own interpretations of the past, at which we have arrived within the context of our own immediate problems."[20]

Like Dewey, Polanyi claims that sheer observation of the world will not lead to scientific success. We *engage* in investigation, commit ourselves, make choices of facts and principles. We never simply confront the world and thereby know it. "Factuality is not science. Only a comparatively few peculiar facts are scientific facts, while the enormous rest are without scientific interest."[21] Both Dewey and Polanyi agree that in knowing we *select* certain facts as relevant. No formal methods will ever overcome the fact that *men* inquire, that they make selections, respond to situations uniquely in some respects. There is a strong tendency to overemphasize the public elements in scientific investigation, to stress the formal features of science. But for Polanyi, "the scientist's procedure is of course methodical. But his methods are but the maxims of an art which he allies in his own original way to the problems of his own choice."[22]

Polanyi suffers from peculiarly rationalistic attitudes, for he claims that we *do* have direct insight into the nature of things and must trust it. The danger that arises in taking seriously the revolutionary human element in science is that apparently no way can be found for appraising the creative and individual contributions of one man by others. Polanyi escapes this difficulty by an appeal to human rationality. We commit ourselves to the quest for truth and are capable of recognizing it when attained. "We should be more frank ... and acknowledge our own faculties for recognizing real entities, the designations of which

[20] *Ibid.*, p. 160.
[21] *Ibid.*, p. 161.
[22] *Ibid.*, p. 311.

form a rational vocabulary."[23] How can a person know and secure a discovery unless he has rational insight into the nature of things? Polanyi fails to see that the openness of science may make truth as a fixed goal unnecessary. This will be discussed later.

In any case, the fact that men *participate* in scientific investigation, that they appraise, feel, apply rules, and make choices – above all that they discover new facts and principles, rules and *ways of knowing* – reveals the poverty of the Doctrine of Fact. Science is a process in which activity occurs, men act, and in which scientific knowledge is produced.

---

[23] *Ibid.*, p. 114.

IV

THE OPEN DIMENSION OF PROCESS

A. *Openness*

Too-casual discussion of process can be devoid of content. We know that events exist in time, come from and move into past and future events. A man is born, lives, and dies. His life, his experience constitute processes. This is so patent a fact that we can ignore it and direct our attention to the particular matters that are of significant concern. Calling life a process points out so obvious a fact that it runs the risk of being totally empty. Everyone knows that processes exist; we need in our particular endeavors at understanding to make distinctions and evolve methods that cut across and into process, revealing important relations and connections. Life includes pervasive and recurring structures – for example, somatic and genetic elements, cellular organization, hormonal dispositions. A biologist who spoke of process rather than biochemical properties would speak vacuously. Knowing what a process is involves knowing its distinctive properties, not the pervasive characteristics it shares with every other thing.

Generic properties of existence can be so obvious as to be unimportant. Yet it is because they are so obvious that they can so easily be neglected. In our quest for knowledge, we move from distinction to distinction, not realizing how far we have come from the commonplace until we find that we have inadvertently denied the transparent and obvious. It is easy to say that knowledge of pervasive attributes is trivial until we find ourselves making claims that contradict it. One of the purposes of analysis is to reveal the ways in which our attempts to gain knowledge of particular phenomena presuppose assumptions that violate quite simple facts.

In this light, the analysis of process often exists in polemic contexts. Clearly process is a pervasive and generic characteristic of things. Any metaphysical description must take this into account. But such descriptions seem trivial until we expose them in their polemic contexts and show how often we neglect the obvious features of process. To ignore the activity involved in science, for example, is to ignore its process-character. A philosophy of science that restricts itself to the study of logical relationships among scientific statements reduces science to but its results. When we point out that science is a process we emphasize that we have omitted something in our analysis. If we emphasize the active elements of knowing we affirm the process-character of science in one dimension, but we may lose ourselves in the flux of change, denying the intelligibility of processes. An arrow flying through the air will strike its target and move through space in a predictable way. There is order within its motion, its path, and its final resting place. Likewise, there are regularities and order within the methods and assumptions of scientific investigation. Science is not a sporadic activity, but a coherent method of rendering existence to our comprehension. To deny this is to reduce science to a blind activity seeking its way out of particular dilemmas, rather than a coherent devotion of human energy to comprehension of the world we live in.

To call something a process is to make an obvious assertion about it, an assertion of such generality as to merit the name "metaphysical," but an assertion that has significant and far-reaching consequences. For it demonstrates the extent to which our prior attempts at understanding have been limited and distorted. The paradoxes and metaphysical dilemmas which arose from ancient conceptions of motion and change rest upon inadequate conceptions of process. The flight of an arrow through the air is a process in time, a process obeying natural laws. Zeno's paradox does not reveal that motion is not a process; it points out certain difficulties in the instruments we have adopted for understanding it. The affirmation that everything is in flux, or the counter that Being is unchanging, both stem from distorted conceptions of the nature of process. Things exist in process and change in time; but they do possess genuine character insofar as we can partially isolate them from the processes of which they are a part. An analysis of the different forms and kinds of processes, and of the important elements of process, may enable us to avoid serious errors.

The ancient claim that everything in the world is in flux is an affirmation of one dimension of process. Mechanically, every material thing is in motion in space, moving from one location to another. As the parts of a thing are augmented and diminished by various complex motions, change of quality occurs. The mechanical model, with its inadequacies, is at least partly a process-model. It reduces all change to motion, but it affirms the transformation of events, the passage of things through time, the development of new arrangements, new objects, new aggregations of material. However, it also affirms permanent characteristics of processes, and antecedently circumscribes and limits the process-character of existence. For while we may expect that insofar as processes are intelligible they are lawful, there is no requirement that the laws they observe be eternal properties of the world. They too may be in process through time, perhaps more slowly. Whether any property of things is fixed and immutable is something for scientific investigation to determine directly and empirically, not to assume without question. Though process by no means entails either the omnipresence of change or its restriction to a specific range of phenomena, it does point out the *possibility* of unrestricted change. Laws are properties and dispositions of particular events; but like all other particular existences they too may be local to our cosmic time and place. Moreover, assertion of such laws in scientific investigations may well demand perpetual reappraisal and reformulation. Process entails the openness of existence and of our knowledge of it.

There are weaker and stronger formulations of the openness of processes. In the weaker sense, when we speak of process we imply that every part or aspect of that process *may* change, *may* be in transformation. We must not close possibilities of change. New facts are discovered, new elements intrude, new kinds of things enter into new kinds of reactions. Processes are open because they extend out in time and space with unfixed boundaries. Any local segment of space-time is significantly related to a wide region of space and time; its boundaries are diffuse and its determination tentative. In Newtonian mechanics mass was a property of matter independent of anything else. A profound possibility was envisaged in the suggestion that mass is a local property of a system dependent in complex ways on other systems, perhaps even a local property of the total distribution of matter in the universe. The process that is an object's passage through space can be viewed as a process inextricably involved with the rest of the material universe.

In the stronger sense, processes always include novel elements; knowledge is restricted; the laws we discover now will prove to be inadequate at a future time. Everything in the world must and will change. Scientific laws are but temporary expedients. Even if they were perfect descriptions of the world at some time, they will be inapplicable in the future. Nothing is eternal; everything is temporary. Likewise, every scientific law is but an attempt to grasp the laws of process, and can only be temporary and inaccurate.[1]

The plausibility of this view derives primarily from an analysis of human experience. The mechanical model seemed to demonstrate the dependence of physical phenomena on a few permanent and unvarying laws and regularities. But the human process, individual or social, is inevitably historical in the sense that it develops uniquely out of a unique past. No two social developments can ever be identical, for their specific antecedents and circumstances are quite different. Social events, in their complexity, occur but *once* and are never precisely repeated. There may be a few disjointed regularities and general tendencies, but these are but a small part of the complex interrelationship that constitutes an actual situation. Only developmental theories of social evolution which view a general trend of society as a development out of its internal properties alone – such as Hegel's and Marx's treatments of human history – may be successful. And they altogether repudiate the conception of knowing human processes in terms of fixed and static categories. The application of specific categories to human experience fails to include the many and diverse influences which can be found within. To imagine predicting the development of human societies in terms of categories derived from their history and that of other societies amounts to a denial that anything new can evolve into the world. Technical and scientific developments are always new; the physical and instrumental possibilities of men constantly increase. New categories, therefore, become necessary for understanding the future, and are recognizable only at a certain stage of development, when the factors they represent have begun their existence.

I do not see any way to choose definitely between the weaker and stronger positions, but I also see little justification for the stronger claim. It is as extreme as the claim that there are definite facts and immutable laws. It is possible to overemphasize the openness of existence and to claim that there are no principles at all for us to know securely, completely denying the possibility of human science.

---

[1] Though mathematical and logical relations seem highly unlikely to change, the above view might well take either of two tacks: (1) that mathematical relations, insofar as they are part of science (synthetic) are subject to the same kind of rejection as was Euclidean geometry; (2) that changelessness is but a characteristic of human willfulness, that mathematics and logic reveal only human decisions to *hold* them fixed. In the latter case, one might argue that logic and mathematics *have* developed and changed, for human intentions are also in process through history. Finally, the consideration of logical and mathematical relations may lead to a qualification of the claim that *everything* will change to the claim that *everything properly located in space and time will*. While relations and laws only *may* change, their formulations *will*, for they always exist in space and at a time.

Science may develop and change unceasingly, but its fruits are either relatively stable and known principles of order, reliable accounts of natural processes, or it is worthless. If there is no definite way in which the results of science fit the world we live in, knowledge is impossible. The stronger position virtually repudiates order in its fascination with change. But *sheer* change is chaos. If the world is intelligible, orderly discoveries can be made. Principles, structures, and laws of development exist.

We must remain content with the weaker formulation. Processes are indefinitely open, but not necessarily in every respect.[2] Any aspect of a process may change because it is connected with so comprehensive a range of things. Life is an adventure; existence is a risk. No matter how secure our knowledge, how well we control particular processes, the future is open. It may introduce something new and unexpected. In the weaker sense all processes, even simple physical ones, involve some degree of risk in their future development.

The weaker sense of the openness of the universe is of great importance. It is the indefinite openness of processes that makes knowledge of any significant extent possible. We live in but a local region of space-time; the evidence we obtain is restricted and limited. It is only because world-processes extend indefinitely in time and space that we can obtain information about remote events. If processes were restricted to narrow regions of existence, then remote events would be forever beyond our grasp. In other words, if knowledge of any significant extent is possible, then it presupposes the indefinite openness of the world.

Perhaps it may be thought that we are unjustified in doubting the existence of at least *some* regularities in the world process. Perhaps there are a few unchanging aspects of things. The world as process need not change in every respect, though it indeed *may*. Perhaps we should weaken our sense of the openness of the universe to the position that many aspects are open and transitory but that some underlying principles of such transformation are immutable. What we have learned about the nature of physical phenomena is that there are secure laws underlying them. We may never fully understand the world, but it is nevertheless lawful and develops according to determinate underlying principles which we may try to comprehend.

---

[2] I regret the vagueness of the word "indefinitely," but it is needed to represent the vague openness of the world we live in. Any given event is indefinitely related to other things and events, but by no means unlimitedly.

Human experience, however, reveals a process that interacts in significant and influential ways with the rest of existence. Consider, albeit unwarrantedly, a narrow range of physical events closed under a set of physical laws. The advent of human knowledge, desires, and abilities transforms the situations in which such laws operate. Human knowledge, which can never be secure, transforms the ways desires are satisfied and abilities are used. The process that is the career of an iron deposit takes on specific properties after discovery and use. Physical laws, when partially formulated and comprehended, produce situations in which desires affect statements of knowledge which affect the possibility of achieving the desires and so on, in ever more complex interactions. Such processes are intelligible in many ways, but they are closed in only trivial respects. If there are permanent physical limitations on human experience, they form but a minor impediment to human capacity and are usually overcome by more or less circuitous methods. From our point of view – and there is no point in trying to speak from any other point of view – every real process is indefinitely open, for human interaction can enter it inadvertently and inexplicably. A routine experiment may be upset by calculated human ingenuity as well as by unexpected discoveries and unsuspected factors. When we call something a process we affirm that it is, from any point of view we are capable of adopting, open to novel influences, unexpected occurrences, even unknown principles. If there are determinate natural laws – and no doubt science testifies at least partially to their existence – they operate in a context of human experience which is unfixed. Given any object known as fully as possible, it can be tampered with and changed. What we know are events, things, *under certain conditions*, processes at a *particular stage of development*. But the conditions may be different, the development at another stage, when we apply what we know in another situation. The generic character of process in the world emphasizes the intrinsic openness of existence to human experience (perhaps by virtue of human experience).

None of this has anything to do with determinism in any sense of that word. The openness of existence as process does not free us from the laws of development which affect us as physical objects or behaving organisms. Scientific knowledge testifies to principles and laws governing human behavior as well as inorganic phenomena. The future of a process may well depend completely on its past, though both the laws and the conditions in which they operate are open from the point of view of human knowledge insofar as they may never be fully determined.

No freedom from lawful behavior exists in a universe open in the sense set forth above. There is only the fact that the future is open, as yet partially unmapped. No human being can possibly know everything insofar as he is part of existence, not outside it. His knowledge transforms his actions and modifies his predictions. So do his desires and rationalizations. A falling object behaves in a certain way if the conditions are appropriate; but a willful man may interfere with the conditions. So may another falling object, a bird, or a strong updraft not taken into account. More knowledge is needed to secure the conditions, but when that knowledge is possessed by elements which affect or are part of the conditions, openness exists inevitably. Our world is a process open in the future, though at the same time it is lawful and intelligible.

All of this Whitehead saw very clearly. On the one hand, "any flux must exhibit the character of internal determination," yet there is also to be found "the final decision of the immediate subject-superject, constituting the ultimate modification of subjective aim ... the final reaction of the self-creative unity of the universe."[3] Even within a causal order, the mental pole of things brings with it what Whitehead calls "reversion": the "origination of conceptual feelings with data which are partially identical with, and partially diverse from, the eternal objects forming the data in the first place."[4] In its subjective phases, an actual entity brings novelty into a determinate order, as human imagination can transcend its physical and sensory origins. Yet this reversion then becomes part of the data for the world's future, marking significant changes in it from the past.

B. Temporality

To call something a process is to testify to its existence in time,[5] to an open future. This suggests that a popular model for the discussion of space-time is dangerous and potentially misleading, though many popular attempts to metricize time in order to apply geometrical analysis to it overlook this fact. It is quite common to overlook the analogical character of the Minkowski space-time diagram (as well as Minkowski's own clear attempt to preserve the uniqueness of the temporal dimen-

---

[3] A. N. Whitehead: *Process and Reality*, New York, 1929, p. 74.

[4] *Ibid.*, p. 40.

[5] The problems which arise in any analysis of time are without end; yet many of them are irrelevant here. We need not concern ourselves with the size of the present, the loss of the past, nor the nonexistence of the future. It is the inevitable and unceasing passage of events and the changes that arise therein that are essential to process.

sion by assigning its tensor components imaginary values), and to treat space-time as a determinate four-dimensional manifold. The Minkowski space-time diagram in which time is represented as a spatial dimension can be informative and useful, so long as it is not conceived too literally. In the literal sense, it suggests that a map can be drawn of the world in time, and on such a map no *process* exists. only a four-dimensional existent. There is no change or development, nor is the future open. There is but a four- instead of a three-dimensional world, fixed in time and space, and determinable by scientific methods. The Minkowskian model supports the conception of a fixed reality to be grasped as we journey along it, rather than an open world in which we act, and which we change in order to understand it.[6]

Yet if we do not take the representation too literally we can see that the Minkowskian world is open too, in both space and time, for while *distant* events are knowable we can never be entirely sure of them either, since they too may be interfered with. To derive the properties of remote events from known laws and observations is no different in principle from predicting (describing) a future event; we must wait for the future for our confirmation (the happening of the event or what confirms it) in any case. The world is open in space as well as time.[7] Travelling in space and collecting evidence are processes also. It is unfortunate that a map of the world, if we do not explicitly maintain its purely representational character and its functional properties, suggests a closed world. If we take such a diagram as a model for reality, we deny the existence of process.

The view that process exists only because *we* exist in time, though

---

[6] The extent to which the structure of scientific theories dominates our metaphysical viewpoint and makes it impossible to see the openness of process can be seen in A. Grünbaum's defense of the quotation on p. 27. To Grünbaum, Weyl does not deny *change*: "there is change in the sense that different kinds of events can (do) occur at different times.... Coming *into* being ("happening"), as contrasted with simply being, is only coming into the present awareness of a sentient organism." (A. Grünbaum: "The Nature of Time," in R. Colodny: *Frontiers of Science and Philosophy*, Pittsburgh, 1962, p. 155.) From the *scientific* point of view everything is, four-dimensionally, and nothing happens or occurs. But three-dimensional entities *do* change, along the temporal dimension. Thus "happening" is, to Grünbaum and Weyl, only a property of what the former calls "the transiency of the Now." (p. 154) It is a "false assumption that 'flux' must be a feature of physical no less than of psychological (common-sense) time." (p. 155)

To Grünbaum it is a unique and scientifically irrelevant fact that (psychologically) things happen to and for us. This is but a property of *consciousness*, not of the objects of consciousness (the world known in science). But as I have shown, existence as process is open in a sense that makes it quite patently false to assert that "everything is," in any sense at all. The Now changes the future; our laying down expectations of the future generally prevents that future from occurring except in trivial and unimportant respects. Grünbaum assumes the being of space-time independent of the Now, denying in fact that the Now creates the future.

[7] A recognition implicit in the theory of relativity.

the universe does not, that our consciousness, ways of knowing, and experience move along a time path which is laid out for understanding – perhaps to a LaPlacean intellect – rests upon the Doctrine of Fact. It neglects the activity of knowing, that knowing is in time, that it changes the world, and therefore itself. It ignores the ways in which a Minkowskian description itself affects the nature of things. The Doctrine of Fact neglects the principle that once facts about the future are known they may cease to be facts, for they can become the source of their own transformation. A Minkowskian representation, if taken literally, makes it impossible to understand human history and social change. History implies the pre-existence of the past; we can diagram it and represent it, but the future is open. There is some plausibility to the analysis of the past developed by George Herbert Mead in which he tries to show that the past too is open, for it is a past understood by us only insofar as we make something of it, utilize and interpret it, not insofar as it has merely happened.[8] In other words, the world is not only open in the future but in the past as well. There is no aspect of things fixed in time, but a world which we investigate, experiment with, and transform. The past is that part of the world rendered to our understanding, and is in constant transformation as we utilize it in the solution of new problems.

The view that the past as well as the future are open to us violates many of our common-sense conceptions of time. It is distinctly an implausible account of historical investigation. In particular, it fails to do justice to the difference between a predictive hypothesis extended into an incomplete future and the evidence given for that hypothesis derived from a completed past. It is true that at different future times we may select different elements of the past as evidence; but nevertheless, as evidence of any sort, the past is available for such use, finished to that extent. Without a past of some definite properties there would be no evidence available for scientific investigation. The past is fixed at least in the sense that it is the source of evidence for hypotheses extended into the open future.

Mead's view, however, does emphasize a very important point – that the world is indefinitely open in almost all respects. This is what is implied in calling it a process. For not only are processes open in time, but in many other respects as well. For example, they are open in that they are never capable of isolation from other systems, and can be transformed by an indefinite number of interactions. It is often tempt-

---

[8] George H. Mead: *The Philosophy of the Present*, Chicago, 1932, *passim*.

ing to conceive of an object, an event, a fact, any natural complex as a *brute* existent, something which can be isolated from other things, studied, and investigated for its essential properties.[9] To some extent this conception has merit, for it is essential to scientific method that experiments be performed in which variables are isolated and objects treated as part of closed systems.

As our knowledge has developed, however, it has revealed that the isolation of factors is but a temporary expedient. The mass of a particular body is bound up with an indefinite range of properties of other bodies. Intermolecular forces exist. Air resistance cannot be neglected except where precision is of minimal importance. Stresses and strains depend on further time derivatives of position. And no force on a particle exists which is quite proportional to its acceleration. In other words, things are never isolated, but extend indefinitely in time and space and interact with remote and unexpected elements. When we say that things are in process, we mean not only that the future is open but that they interact with and are significantly related to diverse and sweeping elements of the world. We cannot know what an object is but by considering it in its associations with other things. We enter it into diverse and manifold processes, interact it with remote elements, and discover its unknown properties. The characteristics of things are not theirs in isolation but belong to them by virtue of the interactions they enter, the diverse processes of which they are a part. Not only is a thing *known* by the interactions in which it plays a role, but ontologically we must conclude that a thing is what it is only by virtue of the various natural complexes with which it participates. A natural complex is everything that it can do, every process it can enter and transform, every interaction it modifies by being a part. We can arbitrarily draw boundaries to any process, to natural complexes, but such boundaries are always open to further investigation and subsequent modification. When we declare human life and experience processes we affirm the possible entrance of novel elements into them. Even simple physical events are processes also, for they are acted upon, influenced and characterized by an indefinite number of factors.

C. The Cumulation of Process

The fact that processes are open should not blind us to their other dimension, that they are determinate in specific ways, however tem-

[9] This is what Whitehead calls the Fallacy of Simple Location. *Process and Reality*, New York, 1929, p. 208.

porarily. All processes are open, but they are not in sheer flux; they are determinate under particular conditions. An object can be isolated from the rest of the world in important and meaningful ways. A process can be developed in momentary cross-section, with significant results. The fact that the world is in process does not render it unintelligible. Insofar as it is intelligible, its constitutive processes possess intelligible characters. The process that is an arrow flying through the air terminates in foreseeable ways; it behaves according to natural laws. As time passes, processes develop new properties; but still into *something*.

In other words, when we call something a process and assert its openness in time and space, we do not repudiate the fact that it is something with a definite character. A temporal process takes on specific characteristics at specific times. We must avoid static categories that deny the change. But we must also avoid the denial that a process possesses distinct properties – the position that it is not knowable because it is in transformation.

Nevertheless, we can make an important distinction between the character a process has in *cross-section*, at a particular time and place, and the character it possesses by virtue of being the kind of process it is. A snowball rolling downhill may be viewed as a total process of movement. It may be viewed in cross-section, at a particular place at a particular time, and of a particular size and weight. But it also possesses properties by virtue of being the process it is. We may emphasize that it grows in size, becomes an object of certain layers and shapes, is subject to certain laws and influences. The process of its rolling downhill is doubly intelligible. There is no reason to assume that any property the snowball takes on is permanent, that such a process will not interact with other processes in subtle and unforeseeable ways. But in being the process it is, it possesses specific properties, and can be concretely specified.

In other words, all processes *cumulate* in the two senses mentioned: they become something, go somewhere; and they possess a determinable character of specifiable functions. We can, if we desire, emphasize the flux of process, the dimension of change. From this point of view, as Heraclitus saw, nothing is. Everything is becoming. Clearly, however, it makes no sense to say that although things are in process, they are never anything. Processes lead somewhere, produce particular if transitory results. A snowball rolling downhill grows in size, becomes interlayered, strikes various things, leaves a mark on the world. From one point of view a process is simply a change in time. From another, the character of a process, though changing, is determinate and specific. If not, knowledge would be impossible.

All our knowledge is an abstraction from processes in which we ignore their interaction with other things. A process is itself open and indefinitely bounded; it never ceases its development. But at every moment, it has an intelligible aspect that is specifiable. Processes can be circumscribed in time or space, and treated as if static and bounded. Knowing demands such restrictive treatment, but is not as such a distortion of things, only a neglect of certain evolutionary features to its own purposes.

The danger in the analysis of process is either to lose the specific properties of a process at a time or in a place in the flux of change; or to neglect its transformative character out of a concern for the intelligibility of concrete aspects of the process. Either result seriously distorts the nature of process. Yet men often attempt to deny the developmental aspect of processes, if not with respect to ordinary phenomena then with respect to the laws underlying such phenomena. On the other hand, in spite of the fact that processes do develop in time, we may indeed analyze their properties. Science is the development of specific instruments for analyzing characteristics of processes, and rendering them available to our control and understanding. We must avoid a view which characterizes a fact as permanent; but it nevertheless is a property of things under definite if not quite specified conditions. We pursue our investigations further, to determine those conditions and to discover what aspects of the process in question we have neglected.

A commitment to process does not, therefore, entail the view that the world is any less intelligible, only that intelligibility does not entail permanent conditions of order. It is sufficient that there be recognizable and determinable properties which enable us to realize our purposes. On the other hand, this commitment does not entail the conclusion that since science too is a process, it is indeterminate in its development, unanalysable. Precisely because it is a process, it may be studied to determine the relations upon which it depends, and the factors essential to its development. Internal structures may be studied and elaborated upon. Above all, we may be able to characterize the general goals and methods of this process upon which rest the knowledge provided by scientific inquiry. For science does lead somewhere, if not to a grasp of underlying truth, then to adequate understanding and comprehension in some other sense. Science is a process, and is therefore something more than any specific set of categories can comprehend. But it is nev-

ertheless specifiable in any relevant respect, to any desired degree. We may study both the logic and psychology of science – though science is something more than either of them alone can reveal. It is science as a complete process, with developing goals and underlying purposes, that is the subject matter proper of the philosophy of science.

v

THE PROCESS OF SCIENCE

A. The Context of Scientific Investigation

Scientific investigation is an individual act; the progress of science rests on individual activity and inspiration. But the sweeping advance of science is rooted in its history. The problems of science arise in a context of theories and experiments derived from previous investigations, developed expectations, and the recalcitrance of existence. The latter must not be overlooked, for it underlies the necessity of empirical validation. The recalcitrance of events, their deviation from expectation, their sheer final arbitrariness, makes confrontation unavoidable. Nevertheless, the theoretical context which gives rise to the particular problem under investigation, and which sets conditions for its solution, determines the ways in which our observations are used.

An individual engaged in scientific investigation is part of a complex situation conditioned by his acceptance of the scientific perspective as well as the characteristics of his environment. He need not inquire and study; he may simply ignore the problems present, or survey the chaos aesthetically. Only if he accepts the conditions set by the perspective of science, utilizes the principles of scientific method and the results of previous discoveries, and examines his present situation as to its interconnections and controls can scientific problems arise. In other words, prior to the existence of scientific problems, scientific methods, goals, purposes, and criteria are presupposed. The stars were visible to many men who never worried about their remoteness, their composition, or their velocities. The notion of an expanding universe and the problem of its origins and rate of expansion are meaningful only in a context of scientific investigation. An ordinary man looks at the sky and sees the beauty or the coldness of the stars. There are no scientific problems there for him. Problems presuppose a framework of education and a devotion to intelligence. Education in science consists both in the impartation of facts and principles, and in the development of skills and goals which

define certain situations as problematic and open to scientific investigation. There is no such thing as an unavoidable problem, nor a bare "fact of the world." There is but a fact accepted in a particular situation to define a problem which is scientifically resolvable; such a problem can exist only in a theoretical and historical context.

In order to overcome the bias and distortion of individual judgment, science must be public and shareable. This is no repudiation of individual contributions in science. Every scientific experiment is performed by a particular individual; every hypothesis is formulated by a particular scientist at a particular time.[1] But these cannot be tested or evaluated except within a public, shareable context. Individual procedures must be appraised and accepted; the context of appraisal is social.

The community of science rests primarily on common conditions and goals which constitute the boundaries of scientific discourse. Men operate within the scientific perspective only when they accept conditions of discourse which arise in a context of antedecently determined principles and facts. A physicist today is governed by the system of Newtonian physics modified for high velocities by relativity theory and for small events by quantum mechanics. A nuclear physicist who denied the discreteness and uncertainty of microscopic phenomena would be puzzling and unintelligible. A scientist who denied the regulative control of logical principles would seem quite bizarre. He would simply not be a scientist.[2]

[1] How easy it is to forget this and to develop procedures that are not only open to public scrutiny, but routine, almost without human participation. The ideal of public verifiability is served best by simple and routine methods that can be applied with a minimum of individual risk. As Polanyi shows (*op. cit.*), every novel contribution to science involves a break with routine and entails individual risk and commitment.

[2] As T. S. Kuhn puts it (*op. cit.*): "Perhaps it is not apparent that a paradigm is a prerequisite to the discovery of laws like these [Boyle's law, Coulomb's law]. We often hear that they are found by examining measurements undertaken for their own sake and without theoretical commitment. But history offers no support for so excessively Baconian a method. Boyle's experiments were not conceivable (and if conceived would have received another interpretation or none at all) until air was recognized as an elastic fluid to which all of the elaborate concepts of hydrostatics could be applied." (p. 28) "One of the things a scientific community acquires with a paradigm is a criterion for choosing problems that, while the paradigm is taken for granted, can be assumed to have solutions. To a great extent these are the only problems that the community will admit as scientific or encourage its members to undertake. Other problems, including many that had previously been standard, are rejected as metaphysical, as the concern of another discipline, or sometimes as just too problematic to be worth the time. A paradigm can, for that matter, even insulate the community from those socially important problems that are not reducible to the puzzle form, because they cannot be stated in terms of the conceptual and instrumental tools the paradigm supplies.... One of the reasons why normal science seems to progress so rapidly is that its practitioners concentrate on problems that only their own lack of ingenuity should keep them from solving." (p. 37)
My references to Kuhn's book are intended as an appeal to his solid scholarship without reference to his very doubtful thesis. Some words, though, should be said concerning the latter. In particular, going beyond the *need* for paradigms in scientific investigation in de-

Science pursues the understanding and control of events. But such an enterprise is carried on in the context of what has gone before, conclusions as well as tried and tested methods, which provide the context for later discovery. New problems which arise are conditioned not only by assumed facts and scientific laws, but by principles of method and form. These are not really "assumed" (as premises), but are nevertheless conditions of investigation, necessary for any scientific work. Such principles constitute the general framework within which science proceeds.

The primary source of the principles which define scientific validity at any time are the reflections and consequences of prior investigations. At different times in the history of science, it was proposed that a gas is composed of point-particles of perfect elasticity; that atoms are miniature solar systems; that quantum particles are very small pieces of ordinary matter. Each of these hypotheses was incorrect in important ways. Yet each of them played a fundamental role in the development of scientific understanding. The first step in formulating a scientific hypothesis is to use whatever information is available, *even if it is not definitely or obviously applicable*. We have no choice in science but to use the results of previous discoveries. Without them we would be in total darkness. Much of what is used in the formulation and development of models is erroneous, but it may be sufficiently correct or stimulating to the imagination to justify its utilization in the solution of problems under consideration. The Bohr model of the hydrogen atom is in many respects inadequate, but it was sufficiently insightful to be adopted with reservations. Despite its errors it promoted sufficient control to allow the discovery of the relationship between subparticles and the frequency of emitted radiation. It encouraged such a discovery despite its inadequacies.

In the consideration of a problem, certain facts are taken for granted, certain principles are viewed as assured. Otherwise there would be no problem. The determination of the velocity of light in space relative to the earth was a problem within the Michelson-Morley experiment only under the hypothesis of an ether. A sophisticated scientific system was presupposed in the very concept of an ether. Not only were Newtonian

---

fining problems of concern, Kuhn affirms that two different paradigms may well be completely incommensurable in that significant problems in one are either trivial or meaningless in the other. If such a view is taken literally, no test exists for determining which of two paradigms is to be adopted on available evidence. While vaguely acknowledging Kuhn's points here, I feel that what he has to say is inexact. Much of the rest of this essay may be veiwed as an attempt to explain what grounds *do* exist for making a significant evaluation in science.

mechanics and optics taken for granted, but also certain preconceptions concerning the possibility of causal action at a distance as well as certain experimental divergences from accepted theories. Einstein's special and general theories of relativity are solutions to problems that arise in the context of Newtonian assumptions, and arise *only* there. The power of the theories of relativity rests on the conceptual revolution they express. Mass and Euclidean geometry ceased to be spatial and physical absolutes. Yet how trivial such a change would be and how unimportant – perhaps not even worthy of scientific interest – if mass and Euclidean space had not previously been conceived to be absolute properties of the universe.[3] The Hegelian view that we move dialectically through opposites to new syntheses may be simply a rendering of the fact that scientists do not really stumble over their conceptual feet, but rather climb a conceptual ladder of their own errors. Every scientific conclusion is in partial error, but an error which makes possible new formulations, which may themselves be erroneous, but which are nevertheless enormously insightful into the nature of things.

The history of science, then, is not something interesting only to the historian, for it reveals the roots of science. Science is a pursuit; it is a human activity, a method implemented and practiced. It includes confirmation and logical analysis, but it also includes wild hypothesization, the building of new guesses out of old material. Unless we believe that the human mind has an innate capacity to recognize structure in nature,[4] we must recognize that valid hypotheses are neither the result of routine methods nor arbitrary investigations. We gain insight in science only by imposing what has been discovered elsewhere on the new problems that confront us. Such an imposition, if careful and meticulous as well as inspired, opens the new area to our understanding and control. Perhaps (almost certainly) our analysis will eventually be rejected. Perhaps knowing is always only partial and erroneous. Science proceeds by proposing solutions amidst an awareness of their ultimate in-

[3] Cf. P. W. Bridgman's claim that we should avoid so defining our concepts and accepting theoretical models that make such revolutionary changes necessary. "We should now make it our business to understand so thoroughly the character of our permanent mental relations to nature that another change in our attitude, such as that due to Einstein, shall be forever impossible." (*The Logic of Modern Physics*, New York, 1928, p. 1.)

[4] Michael Polanyi, as I have indicated, is well aware that routine methods will not suffice for the invention of important hypotheses in scientific inquiry. He is left, then, with nothing but a belief in a rational universe and human capacity to grasp truth. This is because he fails to recognize the process-character of science. The universe is only so rational that we can move from one area to another using analyses that are inadequate in some respects, yet which embody valid and significant insights amidst their inadequacy.

adequacy, by explicitly extending specific formulations to the point where they *must* prove inadequate. All we wish to know is where.

The tendency in logical analysis is to emphasize the rigor of theories, the ground of confirmation. The location of science in social, cultural' and individual contexts is ignored in the pursuit of *objectivity*.[5] Above all, the emphasis on routine logical methods overlooks the sheer audacity that is revealed in supposing, for example, that small pieces of matter (molecules, atoms, and subparticles) have perfect elasticity, occupy no space, that statistical methods presupposing randomness are appropriate. Nothing can justify such assumptions; but they are the heart of the method of science. A theory seeks extension as far as possible in every direction. Often such extensions are in vain. All that is necessary is that they occasionally succeed.

The initial stages of quantum mechanics, for example, were derived from the application of classical concepts of waves and particles despite explicit evidence that they were not adequate to quantum phenomena. Without such conceptual tools no theory could have been developed at all. This is what led Bohr to use classical concepts while relaxing their specificity.[6] The Copenhagen school's defense of the principle of complementarity is rooted in such considerations. We simply cannot invent a new conceptual apparatus out of whole cloth. A wholly new language is meaningless. Science proceeds in an unbreakable historical context. Slowly and painfully it refashions its conceptual systems, and questions what was taken for granted. It can only build on what has come before.

The fact that theoretical construction can never be wholly new, that novel phenomena can only be dealt with by utilizing old and partly invalid theoretical structures, has been pushed to extremes by members of the Copenhagen school in their claim that we can never hope to overcome classical preconceptions.[7] Here we find a new way to close the scientific process – by refusing to entertain the possibility that theoretical invention can overcome any particular historical location.

---

[5] As Polanyi puts it: "the prevailing conception of science seeks – and must seek at all costs – to eliminate from science such passionate, personal, human appraisals of theories, or at least to minimize their function to that of a negligible byplay." (*op. cit.* p. 15.)

[6] N. Bohr: *Atomic Theory and the Description of Nature*, Cambridge, 1932.

[7] "It has sometimes been suggested that one should depart from the classical concepts altogether and that a radical change of the concepts used for describing the experiments might possibly lead to a ... completely objective description of nature. This suggestion, however, rests upon a misunderstanding. . . . Our actual situation in science is such that we *do* use the classical concepts for the description of the experiments. There is no use discussing what could be done if we were other than we are." W. Heisenberg: *Physics and Philosophy*, New York, 1958, p. 56.

It is true that theoretical controls are necessary to the grasp of new events; but once sufficient understanding of these events has been produced, it becomes possible to develop new theoretical orders based on them. Science is rooted in its history but transcends it with profound conceptual revolutions.[8]

There is a confusion here between the stage of scientific discovery at which novel material is digested through older, partially inadequate concepts, and the stage at which this novel material has become well-charted, the inadequacies have been well-defined, and this new domain calls for theoretical invention. Science develops by extending old theories as far as possible; but it then replaces them with new theories of greater adequacy and breadth of application. It is unwarranted to close the scientific process in either direction, to transform science into the routine application of well-defined criteria of method, or to deny that new theoretical visions are possible.

B. *The Conditionality of Science*

Science develops through the extension of older theories into new areas. It may be expected, then, that accepted results of prior investigations, when incorporated into a theory, become the basis for further discovery. They are assumed to be true in order to facilitate further investigation. Under such conditions, many scientific statements tend to lose their empirical character and become presuppositions of these new investigations. Yet, since no scientific statement is quite true, such principles are themselves later found to be inadequate, without entailing the rejection of the discoveries which they made possible. This circular process is the very essence of science.

A striking example of this dynamic process, in which empirical facts become so warranted as to be accepted as true by definition and yet are eventually rejected as false, can be found in the history and development of Newton's three laws of motion. Consider the second law: force equals mass times acceleration. Provided we have independent measures of all three quantities mentioned, we have an empirical fact which can be tested by straightforward experimental procedures. Under limited conditions the behavior of a weight at the end of a spring can be described quite accurately by this law alone. If we take mass to be meas-

---

[8] As Feyerabend puts it: "Should, then, Galileo have tried to get on with the Aristotelian concepts as well as possible because these concepts were the only ones in actual use and as 'there is no use discussing what could be done if we were other (i.e., more ingenious) beings than we are'?" P. K. Feyerabend: "Problems of Microphysics," R. Colodny ed., *op. cit.*, p. 229.

ured by weight and force by extension of the spring, the acceleration and subsequent velocity of the weight can be calculated as can its position at any time. The period and amplitude of vibration can be confirmed by simple observation. Many facts and principles are presupposed in any experiment which is used to confirm this law; but this is natural to scientific inquiry – to accept conditions within which a problem becomes resolvable. On the other hand, when a particle undergoes acceleration, we can use the law to reveal an unsuspected force. The second law then becomes definitional. Similarly, it can be used to define mass under other conditions.[9] It ceases to be an empirical fact *descriptive* of things and becomes a defining principle, a guiding rule for further investigation. In the investigation of electromagnetic phenomena, it is useful to define an unknown force by the acceleration it produces in a specified particle. When a law has been sufficiently warranted it need no longer be treated as an empirical fact, and becomes instead a regulative principle of scientific method.[10] A major conceptual revolution was involved, then, in rejecting it entirely, or at least in recognizing its inadequacy. For with the recognition that mass is not a constant, that force is proportional to induced acceleration only under carefully controlled conditions, the law is shown to be only approximately true under a narrow range of conditions. It is then preposterous that it was ever used to define force and mass.

The movement from empirical fact to regulative principle is fundamental to science as an activity rather than as a timeless mirroring of reality. It is imperative that we accept conditions in a situation in order even to begin to understand it. Such conditions for such a situation are not empirical facts taken for granted. They represent *defining* conditions of the problem under consideration and of the methods to be used in solving them. To "know" in an experiment that force equals mass times acceleration is to recognize that any acceleration represents a force operating. If no force "could" have been present, serious difficulties arise. Our conception of gravitational effects is colored by similar considerations: gravity is a "force," though such a conception is difficult to maintain under presuppositions of causation by contact.

[9] "The definition of mass as the ratio of force to acceleration gained wide acceptance, particularly within the French school of mathematical physicists." M. Jammer: *Concepts of Mass*, New York, 1964, p. 89.

[10] Cf. N. R. Hanson's account of *five* different uses or roles of Newton's second law (*op. cit.*, pp. 100–118). "'the second law of motion,' 'the law of gravitation': these have been construed as the names of discrete propositions. But in physics they are umbrella-titles; they cover everything that '$F = d^2s/dt^2$' or '$F = (Mm)/r^2$' can be used to express – definitions, *a priori* statements, heuristic principles, empirical hypotheses, rules of inference, etc." (p. 112)

Hume's conception of induction, as generalization from past instances, is inappropriate precisely because it overlooks the constant shift in science of the function of statements, from being empirically verifiable to being definitionally presupposed and vice versa. The split between the analytic and the synthetic was permanent and fixed for Hume. The problem of scientific method was to justify generalization from instances to laws; and since such generalization could not be logically justified science was always suspect. When we emphasize the *process* of science, however, we see that generalization in science is rare and usually fruitless. Instead, we have complex hypotheses which are partly confirmed by empirical evidence, and which are then taken for granted in further investigations to extend the range of our problems. The movement from empirical conclusion to formal condition cannot be warranted by ordinary logical principles. It is simply the procedure of science. The results of past investigations are generalized as far as possible in order to set the defining conditions of new investigations. This *is* science; it can be judged only by the future solutions that develop from it. The pursuit of a fully warranted set of criteria for scientific methods is a will-o'-the-wisp; science will lift itself by its own bootstraps if necessary. Perhaps we are fortunate that the world we live in is open to the possibility of rational extrapolation in limited areas, and by further extrapolation capable of indefinite extension in all directions. It need not have been so tractable. But it is far less tractable than Hume supposed.

The conditionality of science further entails that scientific conclusions are never "true" without qualification, that scientific knowledge is unmistakably *hypothetical* in character. This is so in a number of ways. First of all, we never discover natural laws which hold under any and all conditions, but laws of nature which hold under particular circumstances which we cannot define precisely. Laws of science are only valid within specific domains. Relativity theory, classical kinematics, and quantum mechanics all have particular spheres within which they apply. Second, the very form of natural laws is from particular conditions to other conditions. "*If* an object of rest mass M moves with velocity V with respect to observer O, then the mass of the object still with respect to O is $M/\sqrt{1 - V^2/c^2}$." The hypothetical conditions seldom (if ever) can be fully satisfied. "A body free from the influence of external forces will continue to move with constant velocity in a straight line" – such a law is fundamental to science though the hypothesized conditions can never be satisfied. Third, scientific discoveries are found

in contexts that are themselves suspect. "Water boils at 212° F."
assumes not only that a pure sample of water is obtainable but that
the observation of temperature is reliable and well-defined. Such assumptions are necessary; but then our knowledge can never be anything but hypothetical or conditional. No scientific statement is simply true, but is true only under particular conditions. "All men are
mortal" is undeniably valid; but if it is a precise statement, well-confirmed by empirical evidence, it is so only in a particular language
(*this* is a man), and assumes particular principles of investigation and
method. As I will show, determinacy in science can be achieved *only*
by giving up unqualified knowledge. Determinacy and conditionality
are directly related.

On my view, calling science "hypothetical" is no disparagement.
It is science's *strength* to be hypothetical. Its dependence on systematic
organization and theoretical comprehensiveness makes it powerful and
successful in its pursuits; the same dependence renders it hypothetical.
An intelligible world is *presupposed* in science; without it science would
be meaningless; yet such intelligibility cannot be grounded. It is science's presupposition of order that is the heart of its success. Moreover,
its problems of concern, principles of method, criteria of judgment, and
implicit canons of experimental procedure are all virtually defined by
what is presupposed. The very problems that serve to test the laws
and theories under question are given substance by the prevailing
world-view. Without the principles available at a time, *anything* would
be problematic and nothing would then ground success. What is particularly remarkable is that science is, *because* it is conditioned by such
systematic preconceptions (which are not arbitrary inventions, but
themselves based on well-justified presuppositions at another time and
place), successful and capable of precision. The formulation of a generalization or law valid in a limited domain suggests immediately the
extension of the law to other domains to determine if it is valid there as
well. The formulation "under conditions C properties P will hold" is
never completely adequate, for conditions C and properties P are never
capable of completely adequate formulation. But the tentative generalization guides further investigation and compels greater precision
in what is known.

It follows, then, that the logical and theoretical structure of science
is fundamental to it, and that its quest for ever greater theoretical
adequacy and comprehensiveness is rooted in its quest for precision.
Laws are precisely true only when so formulated as to include *all* cir-

cumstances and related properties. Discrete issues in science are but a beginning, setting the problem of determining just how discrete they are. (*Are* mass and color totally unrelated, under any and all conditions?) Science is hypothetical because it continually pursues greater theoretical scope and adequacy, and can do so only by utilizing somewhat inadequate theories as tools. But a consequence of its methods is a continual increase in its comprehensiveness and precision. It is the unhypothetical domains - such as philosophy and art - which can achieve little precision and few definite answers.

C. Logical Order

There are levels of conditionality in the scientific process. Although Newton's second law of motion was implicitly definitional of mass, nevertheless circumstances could force its rejection – when it no longer served its function adequately in the scientific process, when more adequate ways were found of representing the relationship between particles moving at high velocities or over immense distances. Even Euclidean geometry, necessary as it was to science in the 18th and 19th centuries, proved less useful, less valid, under the pressure of new discoveries. It was not a mistake for men to think then that Euclidean geometry set forth conditions *necessary* to a conception of space. They were then necessary. Eighteenth century scientists could not begin to entertain the concept of space apart from Euclidean notions. Space *meant* at that time Euclidean space. Euclidean geometry formed one of the conditions of physical existence, which is fundamentally spatial. It proved ultimately not to be an indispensable condition of science and was eventually replaced. However, it could be replaced only when alternative geometries and physical theories had been developed which cast in a new light the role of geometrical concepts in describing physical space, and when enough was known (based on Euclidean conceptions of space) to raise new questions concerning the validity of such Euclidean conceptions. Euclidean geometry paved the way for its own rejection. Important questions could not have been asked before, for no answer would have been forthcoming. Unanswerable scientific questions are useless if not meaningless.

On the other hand, certain logical principles appear more secure, more basic conditions of science than these. Newtonian laws of motion have been rejected despite attempts to declare them simply conventions or hidden definitions. We may consider the three laws together to be definitional of force, nevertheless, they embody important truths. In fact, they play many roles, for they enter different investigations in different ways. They form differing conditions of the scientific process, depending on the problems under consideration. In some of their roles they are open to empirical disconfirmation. But we do not know how to reject the laws of logic on empirical grounds; indeed, we could have no science if we rejected them.

However, when classical mechanics was dominant it too revealed limits to the ways things could be conceived. How else measure a force but by the acceleration produced? A conceptual revolution was necessary before men could conceive of other possibilities; and so with laws of logic. However, the latter do represent more fundamental conditions of scientific inquiry than do Newton's laws of motion. We would not even consider a set of assertions if they were inconsistent. An illogical science is absurd. On the other hand, logical conditions do not reveal themselves to our rational intuition, nor do they represent limiting conditions of reality, for they can be applied equally to existing and non-existing things. They represent conditions necessary to the scientific (perhaps to any *rational*) process. What is the ground of such conditionality? From the standpoint of science as a process, logical conditions are necessary only insofar as they are conditions *discovered* to be so. They derive their compulsion and power in the movement from empirical discovery to conditionality. It is an empirical discovery that for certain investigations to be successfully completed they must conform to certain conditions. A method *henceforth* will not be considered scientific unless it conforms to specific theoretical and logical conditions. Once such conditions gain logical status, they cannot be rejected empirically: they represent formal properties of order according to which empirical data are categorized. They constitute the abstract conditions of a conceptual system necessary to any scientific theory. They constitute the defining conditions of *evidence*, and science consists in the search for evidence. They are necessary because they are imposed. They are imposed because without some such conditions science would be impossible. The circle cannot be broken.

Insofar as we demand that science validate its results by the collection of evidence, we presuppose criteria for such evaluation. Logical principles are the most powerful of such criteria. They are explicitly *imposed* on the material of science in order that there be investigation at all. Yet they do have a purpose and can be further evaluated by how well they serve that purpose.

A scientist who denies the laws of logic is rightfully thought to be either deluded or joking. If he considers a self-contradictory hypothesis, or even an analytically true one, to be worthy of scientific investigation, he is noted to be "beyond the pale." He cannot be a scientist without accepting conditions of his activities. In a similar, though not quite as forceful a manner, if he claims that Newton's laws of motion are simply and totally false – if, for example, he formulates a theory which contradicts Newton's laws on a macroscopic level – he is a man who does not understand what science is all about. There are men even today who persist in attempting to develop a procedure for trisecting an arbitrary angle with compass and straightedge, or for solving polynomial equations of arbitrary degree by algebraic methods. They do not understand mathematics; they do not understand the material they are working with. Likewise, a scientist must accept logical restrictions on his procedures if he is to be understood by others. He must utilize certain methods, definitions, and conditions. Otherwise he is not a scientist. We imagine that an individual scientist may make a discovery that contradicts even the most hallowed verities of science. The empirical base of science is often construed in this way. However, such a discovery is possible only within a systematic framework. If it attacks the very foundations of that system (as would the claim in the 18th century that the mass of a body varies with its velocity), it is either so compelling that the entire system is modified to conform to it, or it is rejected as meaningless. In either case the discovery is not treated as an empirical fact, but as a conceptual suggestion. For it attacks the conceptual basis of science, not just facts thought forever secure.

Underlying all other conditions of science are the logical principles which form the ground of confirmation. They constitute the framework within which evidence has meaning. Not everything confirms everything else. Confirmation and validation are defined by logical principles. Yet once an evidential framework has been adopted, it is impossible to collect evidence for the logical principles themselves, for they are now at another level of conditionality. They constitute the conditions of evidence and cannot themselves be tested by any evidence that can be discovered. So long as questions arise wholly within the context of these logical principles, they are necessary. Everything discovered must be in accord with them, for they define the terms in which we deal with things. They cannot be false; they can only prove inadequate.

They can be inadequate. In the weaker sense, it is certainly *possible* to find a conceptual framework cumbersome, a hindrance to successful investigation. Perhaps the process-character of existence ultimately demands a less static logic than we now have. Perhaps quantum mechanics demands a three-valued logic.[11] Perhaps a logical system will develop that is quite novel by present standards. There is no reason to assume otherwise. The imperatives within scientific problems will promote their own solutions. It is is hopelessly restrictive to legislate *any* permanent aspects of scientific development.[12]

Such speculations are not wild; there have been conceptual reforms of profound significance in the history of science. Though they have not led to a repudiation of the laws of logic in *some* of their interpretations, they are surely revolutionary in part. In one sense, mathematical logic is an *extension* of classical logic to sentences of other forms as well as to other terms and classes. But such an extension is far more than routine. It entails a reinterpretation of logical order, and the development of new logical principles. These are continuous with the principles of classical logic only in that both are elements necessary for the resolution of problems.

Such extensions are novel logical conditions, of far greater power and control. They arise from new imperatives of mathematics, directed toward new problems. It is absurd to say that they are not *new* logical conditions. They are intimately related to classical logic by virtue of the fact that both constitute conditions of science (rather than conditions of artistic production, moral evaluation, or horsemanship). The development of science revealed the inadequacies of classical logic.

While logical conditions set the terms within which an activity, a goal, or procedure can be said to be scientific and with respect to which they are strictly necessary, such conditions are themselves subject to evaluation or test, not in scientific investigation proper, but in the context of the comprehensive goals of science. We could not even begin to collect evidence unless we possessed criteria which define evidence and its function. Such criteria are necessary to any investigation; they constitute it an investigation. Yet there may be alternative and better modes of conceptualization. Just as Euclidean geometry was abandoned as a a valid conception of spatial relations, so particular logi-

[11] H. Reichenbach: *Philosophic Foundations of Quantum Mechanics*, Berkeley, 1944.

[12] Quine's attack on the analytic-synthetic distinction (*From a Logical Point of View*, Cambridge, Mass., 1953, Ch. 2) is similar to mine. Logical principles and analytic truths are at best only *relatively* more secure than the empirical statements which are tested by the use of these same principles. The whole system can be redesigned so that what was secure against test and appraisal may now be open to empirical confirmation and vice versa. It is the theoretical system as a whole that constitutes significant scientific knowledge, not any particular part of it.

cal principles may prove inadequate, to be replaced by alternative principles. At present we have no idea what these alternatives may be; if we did we would be ready to introduce them. Alternative possibilities usually arise when they can be employed, when they are called for by the existing problems and evidence. Otherwise, like the Democritean theory of atoms, they are arbitrary and unscientific speculations.

Science, as a mode of human activity, is dominated by a conception of what it means to explain events through prediction and control, by collecting evidence for proposed hypotheses, by solving certain kinds of problems. At any point of its development, "understanding" means the exhibition of material in a particular logical form, "prediction" and "control" are set forth in a public manner according to public criteria, and "evidence" refers to data related to hypotheses by logical rules of implication. Otherwise these terms would be empty of significance. But throughout the historical development of science, we find that meanings change, that logical conditions are transformed through the pursuit of an overriding sense of intelligibility and satisfaction. (Vague though this is, it may well be the ultimate ground of any human activity.)

"Satisfaction" is indeed a vague concept; there are many types of satisfaction, within different kinds of events. How recognize satisfaction in science? Most of the time, the criteria of evidence, the conditions of inquiry, are taken for granted and define for us what we call "knowledge." But even they may prove inadequate and may be replaced by alternative conceptions if we find that another conceptual scheme is more satisfactory, reveals more of reality to us, *or*, helps us to solve or avoid certain kinds of problems that our present system is incapable of dealing with. Space *could* be considered Euclidean today, but only if it is recognized that there would then be no way to determine a straight line. When faced with such a problem, we redefine basic terms in our language and reappraise our conceptual apparatus. Space is then seen to possess different properties from what we thought, and we find that we have rejected what was a *necessary* condition of the conception of space. All conditions may follow the same development.

I hesitate to use the word "necessary" here, for since Hume it has been held that the only clear meaning of "necessity" is that of logical necessity. Nevertheless, the point I am making is that the necessity of logical, methodological, and mathematical principles to science is on a par and that they are all (more or less) necessary to science in that they may be said to provide its structure at a particular time. Geometry is necessary to physical science in that without *some* geometry there would exist no representation of space. The very structure of a theory leads from empirical postulates to laws to theories via a pathway of logical and mathematical principles. The latter are not open to empirical disconfirmation, for they define the very *meaning* of disconfirmation. In general, then, they constitute "evidentially necessary" conditions of science within that theory. Such an organization of science reflects the fact that it is usually the nonlogical and nonmathematical postulates that are considered open to test – and rightfully, for if they are found to be incorrect, often only slight modifications of the theory are necessary (such as recognizing the nonconstant nature of mass). But under severe conditions of disconfirmation, the entire structure of the theory may be called into question, and even an entire mathematical system rejected. But it is neither quite correct to say that this rejection implies that that mathematical system was *false* – it was rather no longer adequate ("valid" is a better term); nor that any *given* postulate alone was invalid. The mathematical calculus determines the very structure of the theory and cannot be changed without *another* theory given in its place. Minor transformations of the empirical postulates of the theory permit the retention of the "same" theory. The evidentially necessary postulates – logical, mathematical, and methodological – define the theory and are thus necessary to it, in an extended logical sense. In an older terminology, the "logical" and "mathematical" parts of a theory are *essential* to it; its empirical part may be considered *accidental*.

D. The Principle of Causality

The multiple status of scientific statements, the shift from empirical conclusion to regulative principle, is revealed nowhere so clearly as in the various formulations of the causal principle. The vain attempts of some philosophers to prove that quantum mechanics signifies a breakdown in causality in order to justify a belief in free will, and the equally futile attempts of others to reformulate the causal principle to show that quantum mechanics does not signify any change in it at all, are amazing in their narrow approaches to a complex problem. For what becomes clear is that there is no single causal principle which is strictly either empirical and perhaps disproved by modern developments in physics, or necessary to future scientific progress and therefore not open to empirical disproof of any kind. Causal principles play multiple functions, ranging from the expression of the results of empirical investigations to defining conditions necessary to the entire scientific enterprise. It is merely confusing to select one such function, and legislate it to be *the* meaning of causality.

The principle is sometimes formulated "every event has a cause," which seems to imply that for every event a causal explanation can be found. Put so loosely, the principle has no significant content at all. We must define an acceptable explanation as well as what will be accepted as an event. Surely if an event is nothing less than the state of the universe at a particular time, there indeed exists a prior state of the universe from which it came. But we have eliminated any significant value such a statement can have in science. The problem is then to define the function of the principle of causality so as to render it meaningful. I wish to show that good grounds exist for interpreting this function in a number of different ways.

For example, the statement "every event has a cause" may be interpreted as follows: for any event E there may (in principle) be found a description of some antecedent state of some domain D (perhaps the total universe, though this does not affect this particular formulation) such that in conjunction with the most highly confirmed laws of science a description of E may be inferred, *no matter how precisely E is described*; it is necessary only to make sufficiently accurate measurements of the state of D.[13] This is, of course, only one way of interpreting the causal principle, one which it may be maintained can be found in classical mechanics. The conception of causality contained there seems to imply that if we cannot make perfectly accurate predictions it is because of the inaccuracy of our measuring instruments, not the recalcitrance of the principles of nature, and that as we improve our instruments so the accuracy of our predictions will increase, indefinitely.

This is an *empirical* claim, testable by specific information about prediction in scientific investigations. It is open to confirmation or rejection. We may find, as we sharpen our measuring devices, that a limit of statistical uncertainty is reached beyond which we cannot go, regardless of the accuracy of our initial information.

It is important to realize that classical physics testifies only to the *possibility* that such a principle be true, for a limited range of phenomena.[14] If we wish to affirm the truth of this conception of the principle of

[13] H. Reichenbach: *op. cit.*
[14] As Ernest Nagel points out, classical mechanics in fact implies no more than that a limited number of mechanical states are causally related as described above, but there may exist

causality we must extend it to all possible measurements in a system, not merely those deemed of fundamental importance by the prevalent theory. We may claim that the empirical principle of causality is true, or at least partially confirmed, if the structure of science supports the assumption that given *every* measurement to ideal precision, *every* measurable property of the system is determined with complete precision also. LaPlace's divine intelligence is not limited to mechanical properties, but may predict every determinable property once given complete initial information. Classical mechanics cannot, of course, warrant this extended claim. But it does leave it open as a possibility. It preserves the ideal of complete scientific knowledge of everything in existence. There are no events or properties of physical objects which are not precisely determinable if accurate enough information is given. Statistical regularities, like the fall of a coin, arise only when we do not analyze the system in sufficient detail. So far as classical theory is concerned, if we measure the forces operating in tossing a coin as well as the coin's various physical properties, we can precisely predict its fall. This conception of determinacy is no more than an ideal, but from the standpoint of classical mechanics remained a possibility.

The structure of quantum theory, however, renders quite impossible this ideal of indefinitely accurate prediction. There are certain events which are not rendered more determinate whatever the accuracy of our initial information. We become able to predict with complete accuracy only statistical regularities, such as the fall of a coin over an extended period, but not the occurrence of particular events. If, for example, we could obtain complete initial information (whatever "complete" means, but let it mean anything at all in the framework of science), we would still be unable to predict with precision the time of decay of a specified atom of radium, or the appearance of a particular scintillation on a screen. The laws of quantum mechanics permit only the assignation of a probability to those events, no matter how precisely the antecedent conditions are specified. The empirical principle of causality ceases to be even a possibility for quantum phenomena.

It is important to recognize that this apparent breakdown in the principle of causality does not stem from our inability to measure the simultaneous position and velocity of an electron, nor any other similar departures of quantum theory from classical theory. If we demand that

properties that cannot be analyzed into mechanical properties and which therefore are not causally related according to classical theory. In other words, classical physics is not totally committed to the mechanical model of explanation. E. Nagel: *The Structure of Science*, New York, 1961, Ch. 10.

we be able to determine the exact position and momentum of a quantum particle we are, at least in the context of the present theory, asking for something quite illegitimate; we wish such particles to conform to mechanical laws. But the fact that they do not so conform is not a violation of causality at all. "Waves" and "particles" are concepts of classical physics and apply very loosely in quantum theory. It is far safer to realize that quantum particles do not have positions and momenta in classical terms. Causal anomalies that arise from demanding that electrons obey classical laws are specious. It is, of course, possible to mean by causality "obedience to classical laws of physics." But it then becomes trivial to claim that quantum theory reveals a breakdown in causality. Any new theory also would.

The point is that the empirical conception of causality is not necessarily committed to any particular scientific theory. Even if one causal theory were given up for another, as in the development of the theories of relativity, the new theory might still be able to preserve the sense of completely precise prediction on the basis of the laws of the theory. This empirical principle of causality is preserved as long as the new theory maintains a structure in which precise initial conditions entail precisely determinate events following from them.

It is sometimes claimed that giving up the principle of causality is inadvisable because it states conditions necessary to science – the unceasing pursuit of explanations which may initially resist discovery.[15] It is always possible to redefine causality in order to preserve it, no matter what events transpire. If we did not we would in effect be abandoning the presupposed intelligibility of things which justifies the use of scientific methods. This point has great merit, but it should not lead us to overlook the fact that one or another formulation of the principle has indeed been disconfirmed empirically.

There is an empirical content to the claim that quantum mechanics involves a rejection of causality. We can formulate a conception of determinism sufficiently narrow to have been *disproved* by quantum theory, but which was preserved within classical physics. It is this formulation which was accepted by Einstein as essential to science and which led him to deny the adequacy of quantum theory. The goal of science is to provide completely accurate and determinate explanations of every fully-defined component of physical systems. Quantum theory

---

[15] "The acceptance of the principle of causality as a maxim of inquiry is an *analytic consequence* of what is commonly meant by 'theoretical science.'" E. Nagel: *op cit.*, p. 324.

renders this impossible in principle for all events, for quantum uncertainties affect macroscopic processes.

But the rejection of the empirical principle of causality does not entail a breakdown of science. Scientific knowledge is not drastically curtailed by quantum theory. Rather, quantum mechanics permits significant and valuable predictions which could not otherwise be made. The danger in declaring causality to have been repudiated by modern science lies in the tendency to treat causality as a regulative principle of science rather than as an empirical conclusion. If the principle is necessary to science, its rejection implies the failure of science – a conclusion preposterous to assume from quantum mechanics. The ideals of science have been dealt but a small, not a crushing blow.

The principle of causality is not, in general, treated as an empirical conclusion. It is usually conceived as a regulative principle inherent in scientific aims. To understand a natural event we look for its *causes*, and attempt to find natural laws underlying it. If knowledge is possible it is because the world is determinately structured in some way. The principle of causality expresses the claim that the events of our world may be ordered and understood in some systematic framework. We search for more general laws, for greater comprehensive organization and order. Science is guided in this search by a belief in its own success, that the world is organized in some coherent causal form.

The exact form this principle takes determines what is sought and what is accepted. In the broadest sense we affirm a principle of causality when we attempt to understand anything at all. In this sense the principle is quite unempirical: it is a condition necessary to the perspective of science. Insofar as we believe that we can resolve scientific problems by theoretical inference and experimentation, by conceptualization and analysis, we believe that the world is intelligible. This is a faith in the causal order of things.

It is fruitless to deny the importance of this faith in the possibilities of science. And it is this faith that many men tend to feel is undermined by quantum theory. Clearly, however, this is an erroneous conception, based on arbitrary preconceptions of what form knowledge must take. We tend to demand that knowledge maintain the form it has acquired in past investigations. We like and desire the familiar and are disturbed by change and novelty. Underlying this attitude is the belief that knowledge should be fixed in some ways, not constantly changing in form and structure. Quantum theory is, from this point of view, a violation of our expectations.

This is why it is maintained that the fact that quantum particles are not like classical particles is a violation of causality. Quantum theory violates our demand that scientific theories preserve the same form they have "always" had. A particle has a mass, velocity, and position; so must an electron and neutrino. If not, these are somehow spurious, not fully intelligible. This is affirmed as a limitation on our understanding, expressed as a breakdown in causality. The peculiar error here is that it is definite knowledge that permits us to say that quantum particles are not classical in their properties. We cannot accurately specify both the momenta and positions of quantum particles because they do not possess both simultaneously. They are not the kind of entity that does. But we do know enough about such things to predict their behavior quite accurately and to explain it coherently and systematically. The claim that quantum theory marks a repudiation of causality on this level is an implicit denial that quantum theory deserves to be called "knowledge" – according to some preconceived notion of the form and structure necessary to scientific knowledge.

The point of this discussion is that the principle of causality takes many forms and serves many functions. It is primarily a regulative condition of inquiry, assumed or presupposed in the pursuit of any scientific knowledge at all. In the broadest sense, then, it cannot be rejected except by the complete failure of science. In this form it is even more necessary to science than logical conditions of judgment, for it constitutes the essential faith of the scientific perspective. In narrower formulations, however, it may be taken as a specific expectation of the form and structure of scientific theories and as such may prove illegitimate. Einstein preserved till his death a faith in the truth of the empirical conception of causality, using it as a guiding principle to specific types of explanation. No formulation of the principle of causality is ever purely empirical, but is used to define conditions assumed to be necessary to further inquiry.

Any formulation of the principle of causality serves at least two functions. It represents a faith in certain forms or structures within science, tested by whether science indeed maintains such properties. And in the broadest sense it represents the faith underlying the entire pursuit of science. Both functions are important, and it is foolish to reject either. For new scientific theories do affirm certain possibilites and deny others. Whether we adopt a narrow or broad conception of causality, it is clear that quantum theory does reveal a significant change in some of the characteristics of scientific theories. This does not open the door to free will or to complete chaos; the knowledge contained in quantum theory indicates, in fact, where and how much chaos and randomness actually exist. Certain narrow preconceptions of the structure of science are proven false, but the general faith in ever greater understanding provided by science is preserved. The empirical and necessary characteristics of the various principles of causality must be considered to form a complex network of conditions and conclusions, some of which are illegitimate and unwarranted, others of which are maintained throughout the continuous development of science.

E. The Cumulation of Science

I have considered only one dimension of science as process – that of change and openness. All the conditions, principles, and conclusions of science may prove to be ephemeral and fall before a new theoretical system which provides a better grasp of things. Yet we *do* speak of science providing *knowledge*. How, if nothing is secure and permanent? What can we be said to know if everything may prove untrustworthy? How can we assume that *any* problem is solved, any conclusion warranted, if it rests upon assumptions and conditions that may be questioned and doubted? How can we know anything if it rests on questionable facts, unwarranted principles? Perhaps Hume was correct in saying that nothing can be rationally known. All human knowledge is built on sand, which provides no firm foundation for belief and action.

The answer lies in the way reliable theories are built on false foundations. One example will suffice – the statistical hypothesis in its simplest form, in which molecules are conceived to be perfectly elastic point-particles, devoid of inter-molecular forces, and the associated gas laws describing the behavior of an ideal gas: $PV = nRT$. There is no perfect gas: the gas laws are *false* if interpreted for real gases (though more accurate in some cases than in others). And yet, the kinetic hypothesis provided a theoretical foundation for the gas laws. It appears to be a falsity based on a falsity (for there is no such thing as a perfectly elastic point-particle); yet the ways in which it clearly is inadequate, in ignoring the size and interaction of molecules, also opens the possibility for explaining the deviation of real gases from the ideal. When mass was conceived to be "quantity of matter," every experiment was based on a principle that was quite inadequate (from one point of view). Yet without such a conception, there would have been no meaningful experiments at all. We adopt false or confused hypotheses for simplicity, for convenience, because we have no alternative. We ask questions which can only be answered under the assumption of certain principles, even if these principles are only partly true. Perhaps we discover a more reliable regularity, a principle which is of greater significance, which can be applied to our previous hypothesis to correct and improve it. We move from slightly reliable hypotheses to slightly reliable hypotheses: the path, however, leads to the correction of both until we have replaced them both by more reliable principles.

Processes not only develop and change, but they change into something with its own character and properties. Science as a process is not blind, moving from particular situation to particular situation. This above all is the error of James, Mead, and Dewey. Knowing is a process, but it is cumulative, and what it produces is a conceptual system which provides what we mean when we ask for knowledge of the world. Each step of the scientific process changes into the next stage. Every factor, principle or conclusion is susceptible to rejection. Even the entire conceptual system may be rejected. But science itself grows, and develops a systematic order that includes the significant results of past investigations, yet which is available in new investigations, to solve new problems. If it does not give us a permanent account of the structure of reality, science nevertheless provides us with the equipment to solve problems as they arise. There are questions we can ask that cannot be answered: science develops to the point where it either declares the question scientifically meaningless or answers it. We know our world by being able to act appropriately within it. This is not meant in the narrow sense of physical action alone, but in the sense of any action, whether mental, ideal, or physical. We can build bridges, but we can also draw correct maps of the surface of the moon, and we may someday know the principles underlying the expansion of the universe. Science develops a multiplicity of principles that form a logically coherent system. This system, somewhat like the body of knowledge that is an encyclopedia, can be used to solve diverse and manifold problems. Ideally, we might someday have a scientific system to resolve every legitimate problem that could arise. We would then scientifically know everything. If we isolate the cumulation of science from its development and transformation, we may imagine such a completion; but in fact science can only develop from problem to problem. Each such problem arises from and is dependent upon older solutions; every new solution includes older solutions. It may open new problems, even an infinite number of them, but only if it has successfully solved the problems which gave rise to it. Science is a process that cumulates, and no part of it is permanent; it contains the possibility of rectifying every mistake made in its past, of successfully resolving every problem to which it gives rise. Perhaps, as Kant implied, there are some problems which will never arise in science, since our conceptual system will not permit them to be meaningfully asked. The hypotheses we assume in science may rule out certain possibilities of cognition. Or perhaps there are problems concerning a reality forever outside human experience – concerning supernatural events, unperceived and unperceivable nature, or quantum phenomena before they are observed. Such questions are formulated to transcend the hypothetical nature of science, to break through the theoretical location of scientific inquiry. And science can exist only within a presupposed theoretical context. The claim that there are limits to our scientific understanding makes sense only if we imagine an ultimate reality which we seek to know in some intuitive or nonconceptual way. Science is but one way of *disposing* of things, of predicting, organizing, and regulating events. It seeks decisiveness of resolution and precision of formulation. Other ways of disposing of events and elements of experience that are less precise, less structured, perhaps more comprehensive and gross, may be of greater value than science. They give up determinacy for breadth of application. Science provides a determinate and coherent system of knowledge. It provides us with means for coping with *resolvable* problems that may arise. But it can do so only by dismissing important elements of existence as irrelevant.

F. Scientific Validity

The recognition that science is a process commits us inevitably to a certain duality in our conceptions. Processes cumulate and develop: they change, yet into something determinate. Though nothing can be assumed to be permanent, nevertheless science provides us with a theoretical system which embodies scientific knowledge. There is an unavoidable duality in our understanding of science and its results.

We may ask, for example, if the concept of truth (validity)[16] is meaningful in such a process-conception of science. I maintain that it is, but only by developing a dual conception of validity to match the duality of the process of science.

Since science is an ongoing, developing process, and since all scientific statements are fallible if not invalid, it is difficult to defend the

[16] See above, pp. 31–37.

position that any particular scientific statement is valid in any final sense. Yet a sense can be given to such a notion, if we keep in mind the fact that science has developed its unique conditions. A scientific statement is valid (in science) if it is capable of satisfying the conditions and expectations placed upon it as a scientific statement. A statement is *unrestrictedly valid* in science if it is capable of satisfying every legitimate demand that might be placed on it throughout an indefinite future of scientific investigation. When we assert the unrestricted validity of scientific statements we assert that future investigation *properly carried out* (according to the standards and methods of science) *cannot* lead to its rejection.

Unrestricted validity is the final goal of science, and may be thought of as "scientific truth." A scientific law or theory is "true" if it is now and always will be able to satisfy every proper demand placed upon it. Such a definition, of course, rests on the methods and procedures of science, rather than on the ties between propositions and the world. It reflects the intimate tie between asserting and asserting the truth of a statement, for assertions may properly be understood only in a context of rational investigation within which they function as assertions. A scientific assertion is properly an element of science, and its validity can be determined only through scientific methods. Assertions in ordinary affairs, in philosophy, or in poetry, serve very different functions and must be evaluated within those domains by the standards found there.

The definition above should be contrasted with that of C. S. Peirce: "The opinion which is fated to be ultimately agreed to by all who investigate, is what we mean by the truth, and the object represented in this opinion is the real."[17] First of all, Peirce's reference to the "real" seems to me too judgmental about other rational modes of investigation which may reach conclusions different from though compatible with those of science. The latter's are not the only *real* entities. Much more important, though, is Peirce's reference to the actual future of science in defining truth. For such a formulation is open to the objection that if the world were to be destroyed tomorrow, whatever had been agreed to by scientists up to that moment would be true. Of course, few if any scientific statements are thought to be "true" in any unrestricted sense by practising scientists, so that such an objection loses much of its force. What is presupposed in Peirce's definition is an indefinitely continued rational method of investigation into an imagined future. Surely

---

[17] C. S. Peirce: *op. cit.*, 5.407.

this is both conceivable and possible. However, the same objection – to letting truth depend on what actually happens in science – may be recast by pointing out that it is not beyond the bounds of imagination that the same error should be repeated forever by future scientists, and that they should agree on something false. One reply to this objection – that we would not then know it was false, so there is no point in laboring the issue – confuses knowing that a statement is true with its simply being true, so may be dismissed forthwith. It is quite meaningful that a systematic error should forever exist in science.

There is another reply to this objection, however, which seems to me much more significant. That is that science could not *in principle* (though it could in fact) permit recurrent error. Peirce strongly argues for the self-corrective nature of science, when properly employed. The methods of science involve continual retesting and reevaluation of judgments reached in previous investigations. Errors will be discovered and eliminated. If science were to be unable to eliminate its errors (for example, concerning the remote historical past), or to change so that no attempt was made to do so, it would no longer be science. We would then say that false beliefs were accepted by "so-called" scientists, not that their beliefs were scientifically tested and assured. A simple proper test would show that what they believed was unsatisfactory. Built into Peirce's definition is a sense of science *properly carried out*, rather than erroneous or irrational methods which pass by the name of "science." This, of course, is exactly what I have tried to spell out explicitly, to avoid even an apparent dependence on what the future of science will actually bring.

Three criticisms of the improved formulation must be considered. (1) Any such definition of truth (unrestricted validity) renders it unknowable. I fail to see the force of this criticism. Peirce's definition makes truth no less knowable than does the uncritical conception of truth. We collect evidence to justify the assertion of statements in either case, and there is no reason why this evidence is more properly to be construed as justifying the claim that a statement portrays the facts in some unanalyzed sense, than that it is assertible in any and all properly enacted scientific investigations. A well-confirmed statement is well-confirmed even from Peirce's point of view – though in this case the confirming evidence goes to show that no future scientist *could* find a legitimate reason to reject the statement. "Unknowable" here simply means "without the possibility of error" and *no* empirical statement can be judged so secure.

(2) Historical statements, particularly about the distant and un-documented past, are not capable of being scientifically valid. Precisely! Nor are they invalid. If no evidence *can* be found (in the broadest empirical sense) by which an assertion can be judged, then it is not relevant to science, and neither valid or not. It is dismissed from scientific consideration, not as meaningless, but as not a scientific assertion (though an assertion it may well be in other domains). It is important that the notion of empirical test be construed rather broadly; otherwise mathematics and logic will be dismissed from scientific consideration. But the very strength of the concept of unrestricted validity is to permit the claim that mathematical and logical systems are *valid* in science (if not true or false in some other sense). Euclidean geometry is not valid in relativistic physics, though it cannot be strictly disconfirmed. Scientific validity is a broader notion than that of ordinary truth.

(3) In speaking of "legitimate demands" of science, of "investigations properly carried out," am I not begging the entire question, since what are *legitimate* or *proper* except methods that lead to truth? How else claim that future methods and standards are or are not scientific? The answer can only be – yes, the definition is circular if what is meant is that the determination of unrestricted validity in science depends on some (other) notion of validity; and no, the definition is not circular since the "proper" methods of science are determinined in its own procedures and techniques. A complete reply to this criticism requires a complete analysis of the methods and procedures of science, to be found in the rest of this essay. The methods and standards of science develop under the pressures of tests employed, logical principles espoused, canons of investigation utilized, and other standards of judgment and appraisal – such as fruitfulness, simplicity, and elegance. Not only does science at any particular time possess a fairly definite sense of which particular assertions are acceptable and which are not, but it constantly reappraises the principles utilized in making such judgments according to a comprehensive goal of order, systematic organization, adequacy, and testability. Criteria such as these, implicit or explicit, determine what is unrestrictedly valid in science and also determine the procedures whereby the criteria themselves are amended. What is held valid at a particular time generates new experimental tests, new theoretical hypotheses, even new canons of method which may lead to a redefinition of what is scientifically valid, and so on. The fact that science develops and changes does not vitiate the fact that scientific validity is quite determinable at any specific time. Statements are

unrestrictedly valid in science if their formulation and theoretical role is such that legitimate changes which might (and probably will) take place in science do not render them unsatisfactory. Legitimate changes are those which are proper according to the methods and principles in use at the time under consideration.

This last criticism suffers from a major defect – that of assuming a definite answer to the very problem under consideration. For there is no well-defined "truth" begged in speaking of scientific validity, only a notion brought in unwarrantedly from other contexts where it means something rather different. What is fundamental to scientific validity is the perpetual testing and reexamination which a statement is expected to satisfy if it is to be thought unrestrictedly valid. Such tests are developed only within science itself.

If, however, doubts still remain as to the significance of *unrestricted* validity, then the appeal to the undefined and assumed principles of scientific method may be set aside. The notion of *validity* itself leads to another, less grandiose conception. A statement is not only open to unrestricted demands, but to the specific demands of a particular problem. In particular, a given scientific law may be viewed as applicable and satisfactory within a particular investigation. An inquiry into the behavior of masses in a magnetic field which utilizes a beam scale presupposes the validity (applicability) of Newton's second and third laws. The determination of fever by a mercury thermometer presupposes that laws governing the expansion of metals with heat as well as laws relating temperature and illness apply.

This sense of a statement's validity is always localized in a particular context, within a particular investigation, or under specific conditions. Rather than speaking of the unrestricted validity of scientific statements, and testifying to the future of science, we may speak of their validity under certain conditions, usually without being able fully to specify those conditions. Such statements satisfy the modest demands placed upon them by a particular investigation – they apply sufficiently to render the investigation significant. The law of expansion of metals with heat is not adequate under conditions of extreme temperature if we demand perfect precision of it, but within an experiment in which temperature must be determined to the nearest degree is sufficiently valid to serve the purpose intended.

It must be noted that not every experimental result is valid in this sense. Nor is it entirely proper to speak of a statement's validity "under the particular conditions of that particular investigation," for we may not be able to specify those conditions under which it actually is valid. (Moreover, in science we are seldom interested in a property of a single isolated experiment.) Only well-confirmed statements for which we have satisfactory evidence concerning their reliability under particular conditions may be called *valid under those conditions*. On the other hand, statements do change their status from investigation to investigation. For example, insofar as we accept the complete reliability of perceptual evidence in collecting data in a physical experiment but question it in psychological investigations, we do seem to declare the statement "I see x" valid in one context and doubtful in another. Restricted validity in science refers to the applicability of certain principles within one investigation though open to question elsewhere. The point to keep in mind is that in a particular context the statement under consideration is satisfactory. We must be very sure of its applicability in *this* situation; but it is nevertheless likely that in some other investigations it will fail to be adequate.

On the other hand, we must be very careful to avoid so construing this sense of validity as implying that scientific statements are to be taken as valid *only* for the conditions under which they were formulated. Scientific conclusions as *conclusions* of a particular investigation may, with suitable modesty, be conceived as at best valid for those conditions under which they were confirmed. But they are nevertheless extended into indefinitely remote and complex further investigations. Statements discovered to be valid in one domain are extended into new situations both to test them and to provide a means to test other hypotheses. Every limited validity is potentially unrestrictedly valid. Errors of course arise when statements restrictedly valid are pushed too far, beyond the conditions in which their validity was determined. Yet such extension is fundamental in the development of science. Science works *with* restricted validity, but its goal is complete and unrestricted validity (or truth).

G. *The Particular Sciences*

Although there is an overriding perspective that may be called "science," it is manifested in particular sciences. And although there is a central ideal in every science above and beyond its particular concerns, it is doubtful that there is but a single well-defined scientific method. Particularly, since every investigation is unique in some or many respects, devoted to its own problems and within its own frames of reference, each particular science will vary considerably in its methods and proximate goals. There is little justification for us to assume that precisely the same conditions are present in all investigations, independent of the particular subject matters dealt with.

The view that there is but one scientific *method* rests on the legitimate recognition that there is a comprehensive perspective we take toward things when we attempt to determine their evidential relationships to other things. But it is nevertheless an error to suppose that a single perspective brought to bear on various subject matters in varying contexts, and under different historical conditions, will produce a single method. There is no reason to assume that a given principle of scientific method will prove necessary to any particular science, merely because it has so proved in another.

It is also true that merely because the subject matter of a particular science differs in specifiable respects from the subject matter of another science, is far from sufficient reason for us to forego the attempt to utilize the procedures and criteria of the latter in making our way in the first. The source of the methods and criteria which are presupposed as the conditions of scientific problems may well be the conclusions and principles of method of more advanced sciences. Ways of conceiving problems in order that empirical evidence to resolve them may be forthcoming are developed by each branch of science, and there may be great fruitfulness in utilizing the discoveries of one in the principles of the other. The test resides in the actual problems developed and solutions reached in the specific science in question. It may well be, for example, that rigorous logical criteria are out of place in some of the newer sciences, not because logical principles are alien to such sciences, but because the types of problems that exist *at present* in such sciences are rendered impossible of solution under conditions of severe logical rigor. Even in physics, rigid methodological conditions might well have prevented many great scientific discoveries, by undermining important yet imprecise theories. In the same way, any discussion of analyzing one science in terms of another may prove to be completely without significance if such an analysis does not assist in the solution of important problems in the science under consideration.

In any particular science, investigation is devoted to a stipulated set of problems. Physics, for example, isolates material phenomena from organic interactions and chemical properties and views material bodies from the point of view of Newtonian mechanics and its *extension* into theories of relativity and quantum mechanics. There is no concrete way of distinguishing between physical and chemical properties of physical objects other than by consideration of the problems dealt with and the conditions assumed for their solution. In physics, despite recent advances in electromagnetic, relativistic, and quantum theories, the fundamental concern is for motion of particles, for factors such as position, velocity, force, and mass, and for the continuous phenomena that have been shown to be relevant. Physics expands as new problems arise which demand that expansion. For example, electromagnetic forces produce accelerations of particles. Electric fields produce magnetic forces. Electric fields depend on the resistance of bodies to the flow of electric current, and resistance is dependent on temperature. It is in this manner that physics is extended out of internal concerns and the development of further problems within an accepted theoretical context. Classical physics itself developed out of problems of terrestrial and astronomical motion that were actually of ancient origin. Relativity and quantum theory arose also from such an extension. The reason that chemical properties of bodies are not considered part of the subject matter of physics is that they do not arise in the consideration of its defining problems. There is no other reason why the chemical bond of hydrogen and oxygen in water is not viewed exclusively as a consequence of the operation of physical factors. It is so viewed in physical chemistry – which is an extension of problems of *chemistry* into the domain of physics. Considerable insight into the *problems of chemistry* has been provided by the quantum analysis of chemical bonds. This is what justifies the unification of chemistry and physics in physical chemistry. The problems of concern are chemical, but they are assisted in solution by the application of physical laws. The dividing line between physics and chemistry depends only on the problems of interest. Since everything in the world is material in some respect, physics might be expected in terms of subject matter alone to encompass the entire world of science. But only if such an extension can be justified in terms of the problems of the various sciences involved.

The reason why such an extension does not always take place is not a consequence of the incommensurability of different subject matters, nor of the fact that new or emergent qualities arise in more complex configurations. Rather, extension depends on the discovery of resolvable problems which produce a coherent path from accepted physical theories to biological or social issues. The reason that physics and biology are discrete sciences is that the *problems* with which they are concerned are very different, not because they deal with different spheres of existence. If the concepts of physics *as they are presently conceived* were capable of suitable extension so that they could resolve all the problems of biology – of transmission of genetic characteristics, of miosis, mitosis, neural conduction, and pathological malfunction (and by "could resolve" I mean that they had actually been used in reaching solutions or at least had exposed themselves to clear and acceptable tests) – then there would be only *one* science where formerly there had been two. Reduction is a spurious notion that bridges a real gap between separate and particular sciences.

The reason for the problem of reduction is that we tend to be taken by the view that different sciences exist exclusively because they are devoted to different subject matters. The different and particular sciences exist because, from convenience or historical arbitrariness, men have pursued particular facts in particular areas. There are physical facts, chemical facts, biological facts, psychological facts, and social facts. Nothing is more plausible than that such a categorization is arbitrary and historical in nature. It is far more reasonable that such facts are interrelated, and natural to hypothesize that all the special facts and laws may be "reduced" by appropriate definitions and discoveries to the most general science, physics. In this way the spurious categorization and isolation of particular knowledge would be overcome, and with it the specialization that causes sterile isolation among scientists.

On the other hand, the champion of each of the particular sciences is not so ready to relinquish the significance of his own endeavors and pursuits. The doctrine of reduction sounds to him like a call to abolition; we no longer need biology, for it can be reduced by appropriate definitions to physics – and so also with sociology and psychology. The world must be exhibited as a logical system dependent on the most general laws of physics. To this biologists respond with horror and repugnance: our own specialty is unique, for it has its own unique laws and principles. Reduction is impossible, for *new* properties arise which cannot be predicted on the basis of physical laws alone. Complex configurations of physical entities, when alive, social, or intelligent, take on unique properties. Emergent properties develop. Physics is only one of the many sciences, with its own subject matter and laws. Every science is separate and distinct by virtue of its own subject matter.

Unfortunately, there are no legitimate grounds on which to maintain such a position. The various and separate subject matters of the particular sciences are all part of a coherent and structured world. No doubt complex configurations of matter behave differently from isolated particles; no one would deny it. The view that there are *emer-*

gent properties appears to depend on the presumption of unintelligible and incomprehensible gaps in the world, forever beyond human understanding. Perhaps: but disciplines that begin with such a view have far to go before they can be considered sciences. The primary difficulty with such a view is that it rests upon an inadequate conception of what is involved in reduction – it presumes that the latter would destroy the independent existence of the particular sciences so "reduced" to physics. This simply is not true; while it might be possible, upon the discovery of biological laws relating cellular phenomena to chemical composition, to state the physical laws governing cellular properties, such an analysis presupposes the biological laws as given. The fundamental problems of concern are biological, not physical. Reduction in fact depends upon the existence of known principles and laws in the science to be *reduced* to physics. These are usually to be discovered only in the particular science in question. Its own unique problems and concerns are what define it as a science; without them there would be nothing to gain by reduction. Thus, biology does not vanish under the hypothesis of reduction. On the other hand, successful reduction may well produce the possibility of deriving biological laws from known physical principles. It is incredibly stultifying to decide such matters in advance.

This entire dispute is predicated upon an erroneous conception of the particular sciences. They are not arbitrarily located among particular subject matters. With arbitrary distinctions based exclusively on subject matter, it would be unwonted presumption to legislate the eternal separation of narrow specialities. Our ideal of knowledge is unmistakably that of a coherent system within which all the particular subject matters find a place interrelated with all the rest. However, what must be realized is that reduction depends upon the existence in advance of a well-defined particular science, with its own principles and problems of interest.

The purpose of each particular science is to solve its own particular problems and to explain its own particular phenomena. Until such solutions are well-defined, reduction literally makes no sense. The terms in which an explanation is to be given have not yet been given substance. Science rests upon the selection of certain factors as relevant to a particular investigation – not everything is. The determination of factors relevant to cell division is a matter for biology, not physics. Only when the theoretical structure of biology is so well-defined that demands for further explanation go beyond the bounds of biological phenomena can we seek the physical explanation of the laws of biology. The latter are prior in every sense but that of generality.

Discrete sciences exist, not because men have made arbitrary classifications which developed into vested interests, but because different types of problems arise in the consideration of different subject matters which are not open to the same kinds of analysis. A problem exists only under the presupposition of conditions which define it, and these are adequate only if they not only constitute a question to be answered, but provide sufficient conditions so that evidence may be gathered to answer it. It would be preposterous to maintain that the problem of determining a general law of motion of falling bodies, a law describing viral infection and destruction of a healthy cell, a law stating the factors involved in rapid learning, or finally of ascertaining an adequate explanation of a historic event, are all conditional in the same way, accept the same facts as true, and utilize the same elements in their analysis. Physics developed when motion was revealed to depend on the action of forces on the mass of bodies (sophisticated by moments of inertia, centers of gravity etc.). Viruses are investigated in the context of principles and elements of organic chemistry – amino acids, chemical bonds, the dissolution of cell walls. Human learning may at present be presumed to depend on factors of "native intelligence" – which may itself be a function of biochemical structure, but which is not as yet capable of analysis in these terms – simplicity of task, repetition, frequency of reward, and satisfaction. The problem in question can be conceived only in the context of such elements. It is quite meaningless to ask for the momenta of the masses, the biochemical structures, and electromagnetic fields associated with learning. No problem exists in this framework and certainly none is even remotely resolvable. If in the future we develop our understanding of microscopic processes to the point where we might begin to investigate such problems, science will be faced with new and quite different issues, created by conditions quite different from those of contemporary psychology. Finally, although it is sometimes maintained that history is not a science, if it is one it pursues the resolution of its specific problems in terms of social and economic factors, individual goals and influences, institutional structures and effects, not the relative masses, positions, and velocities of the entities involved. Furthermore, it is by no means clear that we *care* to consider history a science – it may be far more valuable to conceive it as an evaluative discipline, structured explicitly by cultural values and goals. Yet if we restrict ourselves to history as a science, it is clear that whether or not it is possible to explain all historical factors by physical laws is utterly trivial. The significant problems of science concern the specific conditions of scientific investigations and ways of resolving them. It may well be that the future of science will bring such advances in physical knowledge that problems in the special sciences will be transformed. There will be new problems, at that time resolvable, open to the application of the principles of physics. On the other hand, even if a set of differential equations existed in terms of the physical concepts of mass, velocity, and force for everything in existence, but which were not humanly solvable (and there is no antecedent reason to suppose that they would be; many equations in physics are as yet unsolved generally, as those for water turbulence), then reduction would be complete in principle and yet the separate sciences would be still masters of their own domains.

The ideal of a comprehensive theoretical system within which all the sciences are unified coherently is a fundamental goal of science. The desire for explanation in science leads continually to greater systematic order and comprehensiveness. The quest in science is for the most general principles from which all particular facts may be derived. But this should not blind us to the fact that this quest takes place within a framework of *particular* facts which cry out for explanation and particular problems which need solution. A particular science is rooted in its own problems of concern which it seeks to solve. The facts it accepts or presupposes as conditions of its own investigations, the methods it utilizes as valid results of past investigations, the concepts it considers fundamental, even the standards of logical rigor to which it adheres, are its own insofar as it can create resolvable problems in those terms. Only insofar as the results of another science can shed significant light on the problems of a given science can reduction be important or valuable. And it has been valuable in many cases in just this respect. Yet it is foolish to pursue reduction at the expense of the problems of concern in the particular sciences. Nothing can guarantee that reduction will succeed; and what is even more important, its apparent success does not assure us of resolving the problems which are of fundamental concern. Once we realize that each particular science is distinct because of the problems it pursues and the conditions it accepts, it becomes clear that reduction can be a powerful instrument for scientific progress, but must not be blindly sought without regard for its effects on scientific investigations.

The overriding ideal of the scientific perspective seems implicitly to include the hope that all human knowledge may be unified into a single theoretical system. This hope may well underlie the attempt to reduce the particular sciences to a single most general theoretical system. It should be noticed, however, that such unification has not been realized within physics, for quantum and relativity theory are not yet part of a single system; nor may they ever be. We can, with Einstein, only pursue the ideal. What should be realized is that quantum mechanics and relativity theory each have a domain of application within which they provide theoretical order. Gravitational and mechanical problems lead to relativity; chemical and radiation problems to quantum mechanics. It would be a great achievement to unify them in a single theory. Yet it is not an impossible hindrance that such unification is as yet vain. Each theory is significant and valuable in its own domain.

Every particular science has its own problems and phenomena to consider and explain, and must develop its own theoretical system. Whatever its defects, psychoanalytic theory is a genuine attempt to provide theoretical order among complex psychological events. So also with theories of the transmission of disease or the transmission of genetic characteristics. There are no grounds for choosing physical explanation as the only valid explanation except in the ideal sense mentioned above. The key question is just where to look to strengthen biological and psychological theories – in chemistry, or to theories that can be invented without reference to the discoveries of other sciences. It is difficult to know why anyone would think that such questions can be answered in advance. It would seem natural to pursue both physical *and* psychic explanation. Neither would destroy the science of psychology, for it is what provides the explananda of the higher-level theory.

It may be that the problem of reduction exists because only physical science appears to progress swiftly and expeditiously. Perhaps the hope is that by appropriate reduction biology and psychology will be aided in their future development. But whether that is true is yet to be seen, upon the development of appropriately reductive theories, tested by their success. Presently, the particular problems of the special sciences call for particular methods and techniques. Perhaps the desire for reduction stems from the need for theoretical systematization, the lack of which makes it impossible to formulate resolvable problems in the newer sciences. Science progresses only within theoretical contexts constituted by presuppositions of great systematic strength. The social sciences today are often so concerned with emulating other, more successful sciences in their search for logical rigor and empirical assurance that they find it impossible to take any theoretical system for granted. Their theoretical poverty is their great liability. This suggests that any theory – even that of physics – is better than none. Nevertheless, theories can only be accepted if they serve in coping with the problems at issue within the particular science in question. Perhaps if any theory were simply adopted, and utilized as a framework and condition of further investigations until another, more comprehensive theory were found, greater progress would be achieved. But each science must determine its own methods and techniques for itself.

H. Other Domains of Understanding

There are many other modes of experience, many other ways of dealing with the world than through inquiry. Science is but one of many ways of responding to the world around us. Instead of formulating problems which experiment may resolve, instead of examining events to see if our predictions are valid, instead of drawing conclusions from general premises and testing them by observation, we may simply respond to the beauty of things, to their desirability, to the pleasure or pain they produce in us. Science is a unique way of encountering, dealing with, and manipulating things, and it is governed in the largest sense by what we usually call a quest for understanding.

Yet there are other structured and intentional ways of approaching the things of our world which are unscientific and yet which might be called "understanding" them. A artist works methodically and often meticulously. He may not have preconceived, determinate ideas in mind. But he does work serially and purposefully. Each step is a continuation and development of what came before. Things can be made intelligently and methodically; and the outcome of such production can be considered a kind of knowledge: a knowledge of what can be done with certain materials, the kinds of forms they permit, the types of experience a particular product can produce. An artist *knows*, though not in the scientific sense, the results of certain techniques applied to certain activities. He knows the consequences of certain kinds of situations. He is profoundly aware of relationships between potentialities and actualities in some significant area of experience.

An expert at a certain kind of performance – a pianist or tennis player – also knows something about the world in an intimate and profound way. He is aware of the results of certain actions, and though he does not speak about these actions he does respond by further acts, by nonverbal responses. A methodic skill, utilized in certain kinds of situations, may be considered a way of knowing or understanding the world. A pianist responds to certain cues by appropriate and musically profound muscular acts. He performs a written score in an intelligent, skillful manner. He has learned to act with feeling and control in a manner which an equally trained public finds meaningful and important. Such education is a form of knowledge, for it is intelligibly directed, to the intentional realization of desired consequences.

We often distinguish between knowledge *about* and knowledge of *how to do* something. Such a distinction expresses the important difference between knowledge of a skill, an ability to perform according to specific criteria and goals, and verbal knowledge which stems from the demand for conclusions stated and proved. It makes no sense to ask a pianist to *prove* a particular performance, to give evidence for a certain way of performing a particular passage. It is equally futile to demand that a tennis player state his skills in words so that they can be empirically tested. And although an artistic product is open to appreciation and criticism, it is by no means true or false in its assumptions about the world.

John Dewey's theory of knowledge contains within it the possibility of an extended theory of knowledge. In some formulations he appears to say that all intentional activities are solutions to problems and may be considered knowledge. Yet it is not at all clear that art is a form of problem solving; no problem but the very vague one of producing a worthy object can be found. Something is done at each stage of production, but it is more a realization of intended goals and prior steps than a solution to a problem. And it is even stranger to consider a musical performance or a tennis stroke a solution to a problem. Indeed, problems seem to exist meaningfully only in the scientific frame of reference. But in other formulations, Dewey points out that methodic and intentional modes of activity transform the quality of experience, producing more unified, worthwhile, or satisfactory elements in experience. The mark of science or inquiry is that it resolves a problematic situation, and such resolution is recognized by the satisfaction it produces, the unification effected within experience, the increase in value provided. Applied science is more fully or truly science than is theoretical physics because it alone is thoroughly employed within experience to increase and improve its quality. The goal of all intentional activity is the improvement of experience, the increase of satisfaction and diminution of evils.

A broadened conception of knowledge is both interesting and important. There is much to be said for the view that knowledge is both the result and the basis for all intentional and methodic human activity. Science is an activity devoted to the realization of certain goals, utilizing certain procedures and methods. Art and philosophy also utilize particular methods to achieve intended goals. Certain situations ask questions in the sense that they call forth determinate responses, which are themselves open to critical evaluation. Yet such questions need not be verbal, nor need the answers – they may simply produce the desired or expected transformation of experience. Sometimes an act is the only appropriate response. To a poet or painter a sunset *demands* a particular *answer*, and the response is certainly an intelligent, creative act. It makes considerable sense to call this a way of knowing, even if it faces us with problems of distinguishing among different ways of knowing – by stating, doing, or making. It permits us to say that we educate men to *know* their world and to nothing else, for here education becomes the development of all purposeful ways of acting in and upon the things of our world.

I am not interested in developing here an expanded theory of knowledge, but in analyzing the particular process of knowing that is science. For if it is possible to broaden our conception of knowledge as has been suggested, we must determine exactly what distinguishes science from other ways of knowing. For example, it is not possible to define science as knowing by statements, for a poem, a novel, even a manual on piano technique make statements. Stating is one form of human activity] which can be found in different contexts, for different purposes. A statement in a novel may serve aesthetic purposes quite distinct from any concern for confirmation.

Science appears unique in the sense that there is no subject matter to which scientific method cannot be applied – though with varying success. But the view that universal science can preempt all other domains by a single methodic discipline is not properly a claim of science. It is an article of faith of the scientistic religion. Moreover, it errs in supposing that only science is universal: for philosophy, religion, and art are all capable of application to anything. Art seeks no solution to problems, but it manipulates things towards formal and aesthetic goals. Such activity is intelligent and knowledgeable, if not scientific.

The distinguishing feature of scientific statements is the role they play in the search for evidence. In no other methodic discipline is it meaningful to ask for evidence to ground general principles. Implicit in science is a conception of solving problems by collecting evidence. It is

true that evidence is sometimes offered for the solution of quite unscientific problems – for the existence of God, or to ground moral belief. However, science accommodates an ideal of problems for which it is
not only meaningful to ask for evidence, but for which evidence may be
considered to be compelling when obtained. Science investigates problems which are made resolvable by taking well-confirmed principles
for granted, which enable the evidential solution of the problems in
question. Science began with the transition from metaphysical problems which were open to resolution, but not to *determinate* and *decisive* evidence, to problems which in their very conception could be
solved decisively. The demand for a proof of the existence of God *may*
be conceived as a resolvable problem; as such it is scientific. Yet in
many theological contexts, meanings shift throughout which render
evidence irrelevant. No resolution or proof is possible; the issue is
unscientific. Likewise, it is impossible to find contexts in which men
agree as to the nature of evidence which is decisive in the solution of
moral problems. At this stage of analysis, moral problems are unscientific. Evidence can be thought decisive in courts of law, but that is de-
ceptive. As soon as we realize how open legal principles are to interpretation, and to contradictory expert opinions as evidence, we see that
the decisiveness of evidence in law is manufactured. This, however, is
a subject which merits extended discussion, which would not be appropriate here.

It is clear from the account given of the process-character of science
that no problem can be permanently resolvable. This is inherent in the
very nature of a problem. A problem is constituted by defining conditions, logical principles, and assumed facts. As such, it is eminently
resolvable. But in later developments, as the assumed facts and principles are themselves called into question, the original conclusion may
no longer follow. What saves the enterprise from futility is the fact this
conclusion has in the intervening period been confirmed in so many
other contexts that it has itself gained the status of an accepted condition. Science arose when it became clear that under certain conditions
problems were capable of definite resolution. Resolvability is a consequence of the limitation of the scope of problems and the acceptance of a
circumscribed frame of reference. No problem can be solved permanently
in this manner. Resolvability is a property of narrowness of concern,
and particularity of inquiry.

Science presupposes the rejection of certain subjects or methods as
"metaphysical" or "unscientific" – *by definition*. Science is the method

for solving those questions to which there is *an* answer, by the collection of evidence. All other problems, methodic possibilities, are rejected as specious. Yet the quest for answers to philosophic problems – concerning the generic traits of existence, or the nature of time; to religious questions – concerning ultimate loyalties, and the inexplicable; to artistic questions – concerning beautiful forms and novel arrangements; or even to historical and social problems – concerning the importance of events, or the future of social institutions: such answers may not be forthcoming. This does not imply that the quest for such answers is meaningless. On the other hand, without methods of confirmation, such problems are quite unscientific. It is quite possible that we will never be able to correctly predict the structure of future societies; new developments and values which could not have been foreseen may intrude. Yet the scrutiny of future possibilities, however unscientific, is not thereby rendered either foolish or useless. Manifold possibilities must be taken into consideration which make determinacy impossible, but the issues are vital and meaningful. Marx's predictions of future capitalism were false as scientific predictions. But they were powerful, rational visions of social possibilities. Science rejects diversity and multiplicity of outlook; yet it can sometimes be better to retain them at the expense of precision and determinacy.

The concept of evidence requires further analysis. There is great power and value in the quest for evidence and the restriction to resolvable problems. There is a specific compulsion to science as a method. This will now be examined.

EVIDENCE

A. Perspective

There are many ways of acting in the world, many different ways of experiencing. A human individual not only acts sporadically and discretely, but acts in situations constituted by certain orders of existence, governed by certain frames of reference. A given act is what it is by virtue of the governing situation, the order and character it possesses from belonging to a context larger than itself. A given verbal utterance may be either an expression of feeling, an assertion, a command, or a poetic utterance with beauty of form and elegance of expression, depending on the context in which it is uttered. Often we have to refer to the purposes of the speaker in order to distinguish these various possibilities, for the form of an utterance is seldom decisive evidence of its meaning. "The door is open" may be a statement of fact, an expression of a feeling of cold, or a command to shut the door. Indeed, it may be all of them at once. And whether it has one meaning or another does not depend alone on the intent of the speaker, for though he may simply have intended to express a feeling, his audience may respond by closing the door – as to a command.

Everything a human individual does is a human *product*. In the narrower conventional sense a man produces sporadically and intermittently; but in a broader sense he constantly produces: he leaves a mark upon the world consisting of his acts and products. A shout "Fire!" in a crowded theater may be simultaneously an assertion about an event, a command, an exhibition, an injurious action, a crime, or a well-meaning act. The difference among these rests not on the intent of the speaker alone, nor the response of the listeners, but on the total perspective of which they are a part. From a scientific point of view, the speaker is asserting a fact about the theater. In such a perspective facts are asserted, evidence is gathered. To an ordinary person in the theater the word is a suggestion (perhaps with the force of a command) to leave. To a court subsequently evaluating the responsibility for danger and liability, the utterance is a crime, a dangerous action. A sequence of sounds is a composite human product, with quite different characteristics and meanings depending on the perspectives brought to bear.

There are an indefinite number of perspectives governing human production, as many as there are orders of existence giving form to particular products. Even unconscious or unintentional acts operate within physical and social contexts that call them forth and constitute them. A tense social situation may well call forth inadvertent and peculiar gestures from an individual. He may not intend them nor even be aware of what he is saying or doing. But his acts are his products within the given situation: he is responsible for them and they reveal what he is. If he were not the individual he is, if his past had been different, if he had learned to behave in different ways, if the situation were a little different, none of these actions would take place.

Every human product is emitted within certain perspectives and responded to by observers operating in the same or other perspectives. There is no antecedent reason to suppose that these perspectives are the same, nor should they be unless communication is desired. Our man yelling "Fire" may *intend* to convey information only, not to harm anyone or to commit a crime. When a listener responds by desperate acts he has only partially understood the product in the intended perspective. Communication has been slight. There is no particular perspective in which a product must or should be responded to unless communication is desired and important.

I will not analyze here the concept of perspective in general.[1] I only wish to emphasize that such *contexts, orders of existence*, or *frames of reference* (whatever the various illegitimate connotations of each of these phrases) are essential to human production. The word "context" suggests a unique configuration of existence, ordered in a particular way. Yet human perspectives are often reproducible and shareable; we must be able to say that the same perspectives appear in different contexts, under very different circumstances. On the other hand, "frame of reference" tends to suggest something we *refer* to, something we are aware of in our activities and statements. Yet even unconscious and inadvertent production is unmistakably within an order of existence which governs and determines it. Such orders are *not* referred to. Per-

---

[1] For an analysis of the modes of human production and of the concept of perspective in general see Justus Buchler: *Toward a General Theory of Human Judgment*, New York, 1951; and *Nature and Judgment*, New York, 1955.

haps when we act methodically and systematically we try to render perspectives overt and conscious, try to convert them into frames of reference. But they exist whether we are aware of them and refer to them or not. Finally, the word "perspective" tends to suggest point of view, subjective attitude, a distortion of things rather than an actual order of existence within which human production takes place. It would be much safer to speak of "orders of existence," but unfortunately there are orders of existence outside human experience which are not perspectives. None of these formulations can be rendered free of undesirable connotations. I shall, therefore, speak of human perspectives or "orders of experience" in the sense I have tried to set forth, as the orders of existence framing human production giving it its meaning and purpose. Unwanted connotations will simply have to removed as they become hindrances.

B. *The Perspective of Science*

Science is a process, but it is a process governed by certain select perspectives or orders of experience. The process of science is governed by an overriding perspective, for there is an ultimate goal, a conception of meaningful principles and criteria that is essential to being scientific. If we look up at the night sky and ask: "what are those points of light?" we may seek the *scientific* answer to our question if we act appropriately. But we may desire answers in other perspectives, such as: "they are the cold watchers of our fate," or "they are one step higher in the hierarchy of perfection." The question must, therefore, be interpreted in the perspective of science if it is to lead to astronomical investigation.

Insofar as we characterize a particular statement as scientifically valid, a particular method as appropriately scientific, a particular complex of statements, experiments, and goals a science, we assume a coherent order of experience which underlies our appraisals and evaluations. Insofar as a man is educated in science he learns not only experimental techniques, principles of mathematical and logical inference, ways of stating and solving problems, but also a conception of a unified focus of human activity which is dominated throughout by the scientific perspective. The various and distinct sciences are each subperspectives of the overriding perspective of science insofar as they are all *sciences*. The complex order of experience that includes past investigations, future discoveries, accepted methods and techniques, directed into investigative situations, is a perspective.

The central concept of the scientific perspective is the concept of *evidence*.[2] An assertion is deemed part of scientific inquiry only if it falls into a context of searching for and providing evidence. A given assertion may play different roles in this context, for not every scientific statement is open to confirmation by appropriate evidence. It may be taken to be true by definition, or logically true, thereby constituting the ground for other scientific statements; it may be treated as a fact, not requiring further confirmation; or it may be viewed as a hypothesis open to test, to be verified by experiments yet to be performed. Unless it falls somewhere into an evidential framework it is simply unscientific. One of the strengths of the logical analysis of science is that it attempts to provide at a given stage of scientific development answers to the question "what is the relevance of evidence to this (scientific statement)?" An answer must be forthcoming for every significant assertion in science, although different answers will be required at different stages of scientific development.

The ancient hypotheses that everything is water, air, fire, or even atoms were, as then conceived, not part of the scientific perspective. It was then meaningless to ask for *evidence* for such claims, though it was possible and appropriate to ask for *reasons* for holding them. Science and philosophy were at this time indistinguishable because no distinction had yet been made between giving evidence for a claim (or confirming it) and giving reasons for holding it.

The *reasons* given to support the ancient theory of atoms arose from attempts to cope with much broader problems than that of motion. Democritus was concerned with explaining the appearances of things, with reconciling paradoxes of being and nonbeing, and with relating becoming to stable existence. Such a welter of issues could be dealt with only at the sacrifice of assurance and decisiveness. The atomic theory fell into disrepute because its theoretical and philosophic perspective could not produce its confirmation. It was only one of many cosmological theories that could not be chosen decisively among. The contemporary acceptance of an atomic theory is irrelevant here, for the two theories are answers to *different questions*. The greater determinacy of modern atomic theory is the result of its reference to problems so conceived as to yield solutions, conceived after simpler investigations had already taken place.

Evidence can exist only when problems are so well-defined (either

---

[2] As Ernest Nagel says: "the practice of scientific method is the persistent critique of arguments in the light of tried canons for judging the reliability of the procedures by which *evidential* data are obtained, and for assessing the probative force of the *evidence* on which conclusions are based." (*op. cit.*, p. 13) [italics mine.]

because they are circumscribed or because they have grown out of a history of narrow and well-defined problems) that a particular class of observations would effectively rule all alternatives but one out of consideration. The footprint in the sand was evidence to Robinson Crusoe of Friday's existence only within a context of facts taken for granted concerning the types of things which cause indentations in the sand as well as a context in which perceptions are taken at face-value without concern for systematic errors. A theory can be confirmed and accepted only when whole classes of alternatives are ruled out *a priori*. Assumptions concerning the dependence of laws on local time or space, or of the essential rationality of existence, contribute to the background of evidence even when never explicitly stated. The principle of the conservation of matter served as such a regulative principle of chemistry for years, despite its falsity.[3]

Reasons are given under the same fundamental assumptions as the giving of evidence – that existence is rational enough to provide an answer to our questions and that men are rational enough to find it. (I do not consider the appeal to authority the giving of reasons, for it rests on a repudiation of man's intelligence.) The most explicit analysis of the giving of reasons is to be found in Aristotle – we explain an event or a particular thing by stating its teleological antecedents and consequents, by locating it in the world-process. Such explanation, though, need not exhibit any nomological structure; one may locate a particular and describe it without reference to temporal regularities and lawfulness. Evidence, however, exists only when particular laws *entail* certain events within a structure sufficiently determinate as always to entail the *same* events. Evidence exists only within carefully circumscribed limits and arises from the formulation of problems sufficiently narrow in scope to permit decisive resolution. Reasons go beyond such limitations to problems so general in character as to be called "metaphysical," and sometimes to be thought nonempirical. (They are, of course, not nonempirical at all; when we so conceive of the word "empirical" as to mean "scientific," so that "empirical" statements are those to which *evidence* is necessary, then of course we by definition rule out metaphysical statements as nonempirical. What we mean is that they are unscientific. It is a further and fallacious step that leads to the view that only scientific statements are well-formulated.) The teleological structure of Aristotelian explanation permitted the consideration of

---

[3] Properly I should speak here of the regulative function served by such a principle insofar as it was held to be *unrestrictedly valid*. It was later discovered to be only *restrictedly* valid.

normative as well as unnormative phenomena and the unification of these in a comprehensive, if unscientific, system. The greater adequacy of such a system took precedence over the demand for precision and determinacy.

One may always attack reasons given by slightly (though validly) shifting the focus of the problem under consideration. Evidence is given under circumstances such that if the focus of the problem is shifted we may say that we have changed the problem itself. Such circumstances usually depend on the existence of very well-defined investigations or the choice of limited but highly specific problems of concern.

It must not be thought that adequate reasons for a belief cannot be found nor that such reasons do not take observed facts into account. Kant's *Critique of Pure Reason* is a powerful presentation of the nature of knowledge and reality under the hypothesis that apodeictic certainty may actually exist, and it surely rests on common physical facts open to all. Yet it is not appropriate to prove the validity of his particular analysis by collecting evidence for it. There are always other possible ways of approaching the problem – by saying that what Kant takes knowledge to be is not,[4] or that true knowledge is not nomological but intuitive, a becoming with the particular object of knowledge.[5] One's metaphysical point of view plays a role in such issues. Evidence exists only within the metaphysical perspective of science. It exists only when rules of interpretation of a theory are so clear to the entire scientific community that it is possible to test the theory completely and decisively by ruling out alternative possibilities.

Early forms of inquiry were not unscientific in methods chosen nor in their lack of reliance on empirical information, but in the problems considered important. The naturalistic hypotheses of Thales, Anaxagoras, and Democritus were not hypotheses about the scientific nature of things, but were hypotheses which attempted to render the world intelligible in another sense – that of satisfying broad and general demands concerning the relationship between man and his world. The notions of fact and evidence were not relevant, for the questions which the Greeks were interested in were too large and too wide. Evidence arises only after the discovery of fairly simple facts and the presupposition of a theoretical framework of sufficient power to permit the testing and confirmation of proposed hypotheses. It is the

[4] Cf. John Dewey: *The Quest for Certainty*
[5] Cf. Henri Bergson: *Introduction to Metaphysics*

fundamental problem of science to determine these; extrascientific attempts always invert the proper state of affairs, by setting up hypotheses and theoretical principles independent of each other. If we leap too far evidence becomes irrelevant, or at least indecisive. We do not have as yet (and may never have) evidence to settle problems of the origins of the universe nor of our solar system, the moral foundations of human action, or the components of complex human relationships. Such problems are as yet too indefinite in scope and focus. There is no way to determine criteria which are so acceptable to all inquirers as implicitly to define decisive evidence for their solution. It was not until such problems were recognized to be significantly different in such respects from other, more manageable problems, that science gained its unique independence.

An adequate analysis of the concept of evidence is nothing more or less than a complete analysis of the process of science; for what is accepted as evidence at any time derives its status from its location in the scientific process, and the latter changes and develops by modification of its evidential relations. Evidence only exists within a context of accepted logical principles and previous discoveries – that is, evidence is always for or against some hypothesis under certain (other) conditions. The extreme empiricistic view that certain bits of evidence are somehow direct and unquestionable is simply false. It is not sufficient to confirm the statement "the book is on the table" by simply looking and seeing. One must have a very clear idea of the purpose and function of the statement in order that looking be an act of confirmation. May it not be hovering 1/100th of an inch above the table? – not if we know very well in advance that books don't generally do that sort of thing. A theoretical system is always presupposed in even the simplest confirmations. And since theoretical systems are but transitory steps in the scientific process, evidential relationships shift with them. Within the process of science, insofar as it is more than a report of what is commonly accepted, statements shift their evidential status from context to context, investigation to investigation. Logical circularity is an error only in the final stages of science; often the development of science is facilitated by the creation of an intricate network of interconnecting hypotheses and evidence for them in which every statement functions equally well as definitional, hypothetical, or evidential. It is true that evidence is ultimately empirical or experimental, but only in the sense that every acceptable scientific theory must be testable or falsifiable by the results of *some* experiments. The actual logical relationship between the theory and any given experiment may be extremely difficult to determine at any stage other than that in which the theory is actually disconfirmed.

The logical analysis of science has its great value in that it determines the grounds on which we may say that a hypothesis is valid. But it does so in the context of a single set of defining criteria. Insofar as science develops further there are changes in the meaning, function, and status of scientific statements that demand new analyses of evidential relationships. Thus logical analysis can never be sufficient, for it occurs at but one stage of the scientific process. It defines the meaning of evidence at a particular time; but it is therefore forever finding itself inadequate to new developments in science. The fact that evidence can be given only in a well-defined theoretical framework entails that evidential relations constantly be transformed in the scientific process with the theories that structure them.

C. Induction

Most generally, the problem of induction is that of determining the conditions under which scientific assertions may be judged valid on the basis of evidence. The problem of induction is therefore the central problem of science, for induction is the method of science and the justification of induction is the validation of science as a total perspective. Any solution clearly depends upon an adequate analysis of the concept of evidence and the evidential relationship. The entire problem of the nature of validity comes into play here.

Yet the problem of induction is often viewed narrowly and concretely. It is cast in a form determined by present logical and theoretical equipment: we ask under what conditions we are justified in inferring the antecedent from the consequent of an implicatory statement. Indeed, it is often restricted even further, to the problem of validating generalizations by a sub-class of their instances. This further restriction is quite unjustified, and is based on a narrowly empiricistic conception of the greater "truth" of scientific laws. The form in which scientific theories are exhibited is that of a deductive system; we must determine the logical criteria according to which we can maintain high-level hypotheses to be valid when we are capable of experimentally verifying only some of their consequences. Such a formulation of the problem of induction is quite restrictive, since it presupposes a given systematic framework and since it does not consider the problem in any but this narrow context. It is far better to preserve the generality of the problem and to realize that what is at issue is the question of the justification of *any* of the conclusions of science considering that evidence is always particular and local in time and space. Given local and at best alleged "facts," how are unrestricted assertions about remote events to be justified?

Hume's analysis cuts to the heart of the problem by raising the question of whether *any* criteria exist for scientific inference. He demonstrates that such criteria cannot be *logically* necessary, for one can always imagine or conceive a contrary state of affairs, and what is only possible cannot be necessary. If so, in what sense are such criteria justifiable? If we wish to infer a general statement from some particular evidence we must, as Hume saw, presuppose some general principle that holds the past to be like the future in relevant respects. If we knew some principle to the effect that a general law was true whenever a certain body of evidence was known, we could infer from such evidence to laws based on it. Of course, as Hume showed, such a principle could be validated only by assuming itself. Furthermore, though considerable analysis has been devoted to the problem of adequately formulating this principle of the uniformity of nature in order to permit inductive inference, that principle has resisted all such efforts, not merely because it is not itself either demonstrable or confirmable, but because it actually cannot be formulated in such a way as to permit those and only those inferences we hold to be scientifically valid. The key problem is to specify those respects which are *relevant*, for the future is not like the past in all respects, but only in some particular ways. It is the fundamental problem of *science* to determine these; extrascientific attempts only put the cart before the horse.

The attempt to formulate a general principle of induction reflects a desire to render scientific knowledge necessary and science routine in some sense, to achieve the security that deductive methods appear to bring.[6] Yet if Hume's analysis is correct, statements open in any way

[6] Of course, they only appear to do so, for logic and mathematics are by no means closed to inventive contributions. As a purely mathematical system, science would still have room for axiomatic creativity and deductive imagination. It would still be far from routine. The form of mathematics as well as its primary instrument is deductive logical inference, but this does not imply that creative imagination can be minimized in mathematical systems, nor risk eliminated. Rather, imagination must be unleashed to explore new deductive possibilities and to determine the inferential results of postulational alternatives.

However, the inventive contribution of mathematics is quite different from that of science. No doubt new systems of mathematics are derived from individual conceptions of new methods of analysis, new fundamental definitions and methodological principles. New axioms are formulated to produce new deductive systems. But we do not need to choose between such systems mathematically. Any of them is legitimate in its own terms. Diversity is a value in mathematics. In this sense mathematics is a species of art. Science, however, minimizes

cannot be logically validated by a finite set of evidence. Moreover, the Kantian sense of principles of cognition which are permanent features of science does not fit the facts. The rules of evidence are amended and transformed by developments within science itself. When this is recognized, methodological criteria cannot be thought of as necessary in a permanent sense but only as well-justified rules of scientific procedure.[7] Given the methods of science and evidence available, the conclusions to be drawn are straightforward and routine. The criteria themselves, however, may be thought to have been warranted only by their success in past investigations. They may well change as procedures are modified by new developments.

Scientific inferences can never be *demonstrated* to be necessary. The necessity (or compulsion[8]) of a scientific inference is derived from the inquiry-situation in which it takes place, the problem which gives rise to it, and the theoretical system in which the problem is formulated. The problem of induction has been conceived as a search for a final and well-formulated principle expressing the relationship between evidence and explanatory hypotheses. No such general relationship exists. No conclusion follows from any collection of evidence without the granting of many preconceptions, facts, and principles that cannot be well-formulated or explicitly confirmed themselves. What is taken for granted is neither permanent nor is it itself justified but as part of the history and background of the investigation. What we do is ask ourselves specific questions, formulate particular problems, and institute investigations to solve them. The manner in which such questions are asked presupposes an evidential context. Scientific problems are not scientific in character until they have been rendered resolvable by the presupposition of facts and theoretical principles, which permit the determination of a single conclusion based on available evidence. Induction consists in the drawing of a conclusion in a specific situation because of what has been granted. But it is necessary constantly to modify what

diversity and attempts the selection among theoretical systems. No conversion of science into a deductive procedure can escape the need for empirical, and thus contingent, tests. The future may change whatever hopes or expectations we may have.

[7] A further reference to Quine's attack on the distinction between analytic and synthetic statements may be made here (see above, p. 8on). Quine's point is that a decision has to be made in different situations as to what will be the necessary or analytic part of a system and which the contingent or synthetic. No final decision can be made, for the analytic part of a language may be transformed in various ways to suit different purposes. Only the system as a whole may be considered under definite empirical test. Thus even methodological criteria and definitions may be thought of as justified by their pragmatic function in scientific investigation.

[8] See below.

is presupposed; the theoretical contest of science shifts constantly.

It is unquestionable that within any science at a particular time there exist implicit and explicit methods and criteria of analysis. Without such there would be no science at all. But when science moves on to new areas it can do so only by rejecting some very common principles that had not been thought before to be so questionable. It is tempting and dangerous then to suppose that serious logical errors have been committed.

For example, the unification by Newton of terrestrial and astronomical phenomena was a violation of the medieval doctrine of the separation of these two realms of being as sources of evidence. The theory of relativity violated the demand that space be conceived as Euclidean in order to preserve continuity with what is imaginable. The Copenhagen interpretation of quantum phenomena has been criticized for violating the demand that science preserve descriptions of the unknown in terms of what is known – it violates our sense of *events* to speak of quantum *particles* which do not possess both momentum and velocity simultaneously – though quantum mechanics may on the other hand be thought to have conclusively shown that quantum phenomena are the result of events that are in such respects quite different from macroscopic events. Such principles *must* be violated in the development of science; and it is therefore unwise to suppose that principles *necessary* to science can be adequately formulated in any but a very humble and tentative sense.

Perhaps it may be thought that we do not really need any justification of induction.[9] We have no alternative *but* to depend on past experience in science. All evidence is past evidence; we cannot find an escape from this fact. We cannot ask for a justification of induction because that is just the way we know. It is not rational to reject induction; it is therefore not rational to seek to justify it. It is absurd to attempt to justify it because it is necessary to scientific investigation. It is the very method by which science has been successful. If science is a process it makes no sense whatsoever to ask for the justification of scientific theories from past experience. Science as a process must depend on evidence antecedently available. Not only is there risk present in holding a particular hypothesis which may turn out to be false but there is always risk involved in asserting any particular fact to be a

---

[9] Cf. Norman Campbell: "our 'knowledge' of future events simply *is* something based on our knowledge of past events.... it is absurd to ask why it is based on past experience." (*op. cit.*, p. 60.)

fact. The empirical dimension of science develops from the truth that no scientific assertion can be verified in any complete sense. Evidence is needed, evidence obtained from methodic and careful experimentation, *performed in time*. And evidence can never fully warrant a scientific statement; it can only validate it as a conclusion valid under the conditions of particular circumstances.

This conception of the temporality of the scientific process, though insightful into some of the fundamental properties of that process, is nevertheless too loose, for science is not alone in depending on its past. Philosophy and art also grow out of past traditions and earlier formulations. The problem of induction concerns the validation of scientific conclusions insofar as it differs from validation in other perspectives. The problem of induction is not that of verifying a scientific assertion for all time, nor of *proving* a general conclusion from particular data, but that of specifying what it means to *validate* a scientific assertion under the condition that it may (indeed, probably *will*) prove inadequate in later investigations. In what does such tentative validity consist, and why should we suppose that such a doubtful and insecure procedure provides us with the truth about natural phenomena?

It is clear that such questions concern nothing but the nature of evidence – the evidential relationship and its development in the process of science. Hume was correct that the problem of induction is fundamental to science, because it essentially cuts to the heart of the evidential relationship. In fact, it is not possible adequately to explain the nature of evidence in science without discussing both the ontology of science and its development as a process. This entire essay can be viewed as an attempt to analyze inductive procedures as the exhibition of the evidential relation underlying science.

The view that science requires the justification of inductive procedures in some final sense errs in supposing that such justification can be given apart from consideration of science as a total enterprise. Science does not employ methods it is possible to justify independently of itself; it rather is the very process in which scientifically rational methods are discovered and elaborated.[10] Inductive procedures are the methods which have served in the successful resolution of scientific problems. They certainly *may* cease to be successful tomorrow; that is the inescapable risk of all knowing. We cannot ask for a *proof* that the

[10] Strawson's "solution" to the problem of induction expresses a position similar to the one above, but errs it seems to me in looking to "ordinary" contexts of justification. P. F. Strawson: *Introduction to Logical Theory*, New York, 1952, Ch. 9.

future will be securely like the past, partly because it indeed will not be, but primarily because such a proof would be possible only in a closed world! Evidence is always past, yet knowing is always into an open future. This is a problem only if we wish implicitly to overcome the openness of that future. If *past* evidence does not justify our making predictions about future events, what could? Only evidence from a future that is in some sense (since it is already fully known) not a genuine future. The future is in fact open; natural events are irremediably contingent; scientific methods can only move from an assured past to a hypothetical future. The gap between them can never be bridged without risk.

Validity in science stems from the nature of the problems dealt with and the conditions under which they are formulated. One of Hume's problems – which he never really solves – is to explain how a single experiment can confirm a scientific hypothesis. Such confirmation is possible only when the problem in which the hypothesis is formulated creates conditions which permit such a validation. If, for example, we wish to test the hardness of a new alloy, we drop a standard diamond weight on it and measure the height of the rebound. We *suppose* that such a measure is useful, that different samples of this alloy will prove homogeneous with respect to this property, that the deformation of a softer substance will have particular kinetic results. The conclusion is meaningful and valid only under such assumptions. Our simplest and most direct observational measurement would be meaningless if we could not be sure of a certain degree of stability in our measuring instruments.

Science devotes itself to the search for resolvable problems; certain methods reveal themselves to be more successful in producing these than others. Under the assumption of certain conditions, some questions can be answered in a straightforward fashion by suitable experiments. For example, if the conditions of an experiment are all highly warranted, in many and diverse instances, we may question the validity of a relatively new and therefore insecure hypothesis. Let us suppose that under these conditions it is disconfirmed. Let us also suppose that under other, equally acceptable conditions it proves valid. A new problem is revealed in which some conditions, highly warranted in past investigations, are now open to question. We are not sure which, and further investigation is necessary to inform us. But our further inquiry depends upon the context we have taken for granted. There are an infinite number of hypotheses which might reconcile the difficulty (involving special forces, temporal variables etc.). The solution which is accepted depends on the assumptions made, either explicitly or implicitly. The problem is actually meaningless apart from the conditionality of our investigation.

On the other hand, we must also recognize that science proceeds by the extension of validity as far as possible in new investigations. Every scientific investigation is particular and leads to conclusions that are at best conditionally validated. No scientific hypothesis can be absolutely confirmed; it can be confirmed only tentatively, under particular assumptions and conditions. Scientific confirmation is thus *conditional*. But it is clear that constant avowal in subsequent investigations of this conditionality would be a serious limitation. It would not be possible to formulate resolvable problems if no facts could be assumed without qualification. We therefore *assume* that validity may be extended without limit, or at least as far as is necessary for the pursuit of other investigations. Such an extension is always partially unjustified – this is what Hume realized. But he felt this to be a limitation of science. Rather, it is the principle which produces scientific progress and development. There is always risk in such an extension, but this risk is necessary for it permits the development of new problems which in such contexts are themselves resolvable.

Science always takes conclusions which are only restrictedly validated and extends them as far as possible in all directions. Every scientific confirmation is limited and conditional, but the primary movement in science is from such conditioned or hypothetical conclusions to necessary conditions of subsequent investigations. A partially validated theory may be used to define other concepts, to constitute the framework of new problems. In this way science develops as a system, open to further tests as a whole. There is no logical reason why every scientific conclusion is not emphasized to be limited and hypothetical. A conceptual revolution such as that produced by the theories of relativity or quantum mechanics would not then be a revolution but simply a recognition of expected limitations of outmoded theories. Only when science develops as a system within which particular investigations receive their meaning and direction is a revolution necessary. It is a revolution because it repudiates assumptions that have become *necessary* to science. It takes place because theories have been extended far beyond their grounds of confirmation. Yet such extension, though unconfirmed, is the fundamental step in science. Risk enters into science because partially warranted assumptions have been extended too far. Yet if they were not so extended, scientific progress would be much less swift if it did not cease entirely.

The systematic organization of the conclusions of science is essential because it forms a conditional context within which highly sophisticated and detailed questions can be asked and answered. In the development of science theories of great richness and power are proposed and accepted as validated solutions to well-known problems. When these are developed systematically, considerable scientific progress can be achieved quite straightforwardly within the context of their acceptance. The reason laws and theories known to be invalid in some circumstances are so essential to science is that they too possess systematic power. They permit systematic investigation and the drawing of consequences, which may later be corrected without difficulty in other investigations. Such systematic power is far more important to science than the "truth" of scientific theories. Through it theories become conditions of subsequent investigations and lead to new hypothetical conclusions. Without such conditions there could be no inquiry at all. Invalidity which can later be adjusted is but a small price to pay for the entire future development of science.

The inductive criteria of science are those fundamental principles of method which cannot in a given investigation be questioned – they indeed are those principles which make the investigation possible at all. They cannot be questioned, for they are presupposed in the very formulation of the problem. Yet there is no unqualified sense in which they provide validity. They do so only insofar as other investigations in which they were used have provided successful results *by the very standards they define*. This is a circular process, but it is not a vicious circle, for the constant shift in science from empirical generalization to defining condition provides a constant source of checks and balances. We can test a given hypothesis only under certain conditions taken for granted; but we can test any of the conditions themselves by taking something else for granted elsewhere. This continual process, forever insecure and full of risk, is science.

D. The Compulsion of Science

The perspective of science may be defined by the presence within it of *evidence*. What I shall consider now is not the nature of evidence so much as the character of a perspective dependent on the collection of evidence. We are often told that no statement is rational unless it has been supported (or at least is supportable) by evidence. The methods of science are the only rational methods. In some sense or other we are compelled by evidence and reason.

There are perspectives in which we are not at all interested in securing belief, and instead seek aesthetic satisfaction, sensual pleasures, objects made in exciting ways. There are many ways of dealing with things, and only a few have anything to do with grounding belief. Moreover, of the latter only one way (complexly viewed) can be said to be scientific. In this we recognize not only the relevance of the testimony of others, the possibility of being wrong, but agree that it is meaningful and necessary to seek evidence for our conclusions. When we search for and give evidence for a proposed hypothesis, we have adopted the perspective of science.

Implicit in the central role of evidence in science is a unique compulsion, one which underlies the conception of science as the only rational or justifiable method. This is partly due to the extension of narrow conceptions of scientific method into inappropriate areas; but there is a prevailing mystique as well that science alone provides control and understanding. The perspective of science is compelling in some definite way.

We cannot claim without further qualification that scientific methods are the only methods which provide us with knowledge. "Knowledge" is not a term which denotes a perfect grasp of reality; not even science can provide that. Science is one of many ways of controlling and directing human experience, but it is by no means the only one. Science provides evidence for the solution of particular problems. But this may be considered the best way to obtain knowledge only if we so *define* knowledge. There have always been men who seriously doubted this. Perhaps divine inspiration is more profound and illuminating if somewhat unreliable. Science rejects as irrelevant to its procedures many important questions – concerning the nature of God and the mysterious, the source of moral judgments, the foundation of values. It does so by simply declaring such issues unscientific. It is no wonder that poets, artists, even ordinary men are somewhat sceptical of the value of science. If science is seductive and compelling, it may be a considerable danger to important values. Philosophies of science which declare all unscientific problems to be meaningless are even more dangerous, for they extend the model of science everywhere, and hold what is not amenable to this to be outside the domain of cognition.

Social and psychological compulsion are peripheral to this essay. There are, no doubt significant reasons why the perpective of science is today compelling to those who attempt to extend its purview in all directions, who are so taken by its powers that they deny the possibility of other methodic activity. The perspective of science has led to profoundly important technological and social changes; we are more and more a culture given its shape by technological applications of scientific research. The fruits of scientific study are real and tangible; they produce a sense of powerful labor. A gifted man, capable of work in a number of areas, chooses science to make his efforts obviously productive and efficacious. Due to the ready achievements of science, he overcomes intellectual alienation and comes into productive relationship with his world. Theoretical speculations in younger sciences are dismissed when their results are not tangible and evident. Tangibility is the mark of satisfactory and rewarding work. The scientific method is labor with the greatest probability of success. With its immediately apparent results and its powerful social implications, science virtually guarantees some reward to its practitioners.

Psychologically science is even more appealing, because it deals with *resolvable* problems. It overcomes the frustration of struggle with open problems which do not have straightforward, determinate solutions by routinely dismissing such problems as alien. Santayana has noted that no single philosophy can be offered for the understanding of human experience – multiplicity of philosophic viewpoint is not only inevitable but valuable.[11] Philosophy offers itself in diverse and multifarious forms; no final philosophy has been forthcoming, nor have those offered been compelling so determinately as solutions to scientific problems. Similarly, there is no one way of writing a poem nor of creating an important painting. In fact, there is no single way of performing scientifically; but a scientific assertion can be viewed as "true" in the sense that it is, in the context of the problem in which it arises, the only one acceptable. A scientific conclusion, in other words, in being warranted by available evidence, must be acceptable to everyone who understands the issues. No similar unanimity is necessary nor possible with regard to a poem or a philosophic system. Multiplicity and diversity are essential to art and philosophy; but they are legitimate only in quite restricted areas of scientific investigation. The fact that science deals with resolvable problems renders it a far more secure activity than art or philosophy. Of course, if it becomes *too* determinate so that individual initiative is suppressed and routine methods dictate all conclusions, progress in science may cease; but when scientific adventure

---

[11] G. Santayana: *Obiter Scripta*, New York, 1936, pp. 94–107.

and initiative are emphasized, the goal in science is ultimately for *suffi-cient* evidence for a hypothesis or theory. The very notion of sufficiency reveals how science differs from other enterprises – how it is uniquely compelling.[12]

E. The Compulsion of Evidence

The primary sense in which science is compelling must be distin-guished from the psychological and social factors which produce a particular mode of belief. There are times when we say that only one mode of activity can be performed at a particular time, when an indi-vidual can choose from only a certain range of events. In the most ob-vious sense, a man is compelled by his physical environment to act in a limited number of ways. We must eat and breathe if we are to live. We must remain on the surface of the earth unless we use mechanical equipment. The story of Icarus reveals the intense awareness the Greeks had of such compulsions.

Similar compulsions govern the choices men make, sharply determi-ning the options available to them. Such compulsions lead a football quarterback to choose this particular play rather than another, to run rather than to pass; an artist to draw this line rather than another; a

---

[12] Jacques Barzun, in *Science, The Glorious Entertainment*, New York, 1964, delivers a crisp and forthright, if not always accurate, picture of the social and psychological elements in our scientistic culture. "It is one of the great advantages of the scientific method, as Bacon long ago pointed out, that it can raise ordinary ability above what might be expected of it: science is the democratic technique par excellence." (p. 75) "Our strongest faith is that there is no limit to the good and the useful, no drawbacks to a predicted success. By trying, trying scientifically, we shall end by reaching any goal we conceive." (p. 70) Barzun cannot appre-ciate the virtues of such democratic methods. Yet there is considerable truth in his analysis of the consequences of scientism. "The mass output of small 'contributions' results in the great paradox of abundance: our culture is enormously productive of knowledge as of goods. We know – as the saying goes – more than ever before. But who is 'we'? And *where* in a prac-tical sense is the knowledge that our many independent groups of specialists produce?" (p. 130) "Only a scientific culture tends to require that all knowledge take but one form." (p. 196) "The desire of all scholars to emulate physical science ... has been dearly paid for.... 'If research is the right name for good work in Science, then it is the wrong name for academic scholarship, good or bad.'" (p. 129) "In truth, the commandments derived from science and techne and clustering in the institution of research boil down to a diffuse, pervasive fear – fear of fellow professionals, fear of one's own mind and ego as 'subjective,' fear of errors creep-ing into a text while one's back is turned. The resulting temper is not, on the face of it, favor-able to good work. Indeed, what fear inspires is work other than that which might spontane-ously attract the mind; the job must fit the means and curiosity be redirected to questions ('problems') that can be managed without running the prohibited risks. The mind, in short, is subdued by the mechanical." (p. 137) "Ours is the first time in history that a man has been *expected* to be inventive – and by means of *research*." (p. 140) Although Barzun's sense of the role of art is really nothing more than a glorification of mystery as profound, the insufficiency of science comes through very clearly. "If man is to enjoy on earth the emotions proper to a native, provision must be made both for feeling at home and for contemplating mystery. The two are not incompatible; any great object of contemplation (science included) can give rise to each. The difficulty is to enlighten these emotions equally." (p. 229)

mathematician to infer but one conclusion from a particular set of axioms at a particular time. A man addressing an audience, interested in persuading them, must utilize specific techniques, rhetorical phrases, a language they understand. A chess player, after a short opening, must develop his pieces in only some of many possible ways or he will lose. A physicist, setting up an experiment to resolve a particular problem, must deploy his material and forces in definite ways or his experiment will be worthless. There is a sharply limited range of valid activities in each of these cases, restricted by the situations and perspectives in operation. Compulsions exist *within* these perspectives. The speaker need not try to persuade, the chess player need not try to win. But if they do, they are compelled to act in highly specific ways.

On the other hand, they are never completely compelled to act in any particular manner by a given perspective. There are alternative methods of persuasion, different words that might be chosen, different examples and modes of address. Chess demands versatility and flexibility as well as organized and systematic play. There are alternative ways of setting up an experiment, arranging the materials to produce the same experimental results. Men eat and breathe in different ways. Wherever there are compulsions there are options as well; within any perspective there are elements compelled by the situation and others open to alternatives. An individual functioning in a certain perspective must act in particular ways if he wishes to remain within that perspective. But he nevertheless always has options or choices open to him, for no perspective can limit action completely. Something else might always have been done within that same perspective.

The physical elements of a situation do not compel a given type of activity; it is the perspective that produces whatever compulsions exist. In the same physical situation, but from another point of view, such options and compulsions may be reversed. A description of the types of compulsion and option that exist, and the situations in which they are called forth, is probably the most telling analysis it is possible to give of a perspective. To describe what is involved in artistic production, the type of intelligence required, the methods employed, is to describe the compulsions present and options open. To describe the scientific perspective is to tell the acts that are necessary, the circumstances under which they hold, and the alternatives that remain.

The concept of evidence is fundamental to the scientific perspective. Yet the concept of evidence is not clear or self-sufficient apart from the methods and techniques utilized in scientific practice. Every element of the scientific process plays some role in the quest for evidence, but it may play different roles at different times, and need not itself be open to evidential confirmation in any given investigation. The basic aim, however, and the source of the underlying criteria by which the scientific process develops, is for the determinate solution to problems as they arise by the collection of appropriate evidence. Only those problems which can be resolved by virtue of the way they are conceived and conditioned are scientific. Only those hypotheses which in principle can be tested determinately against alternatives are relevant to science. In other words, science searches for methods and techniques whereby its conclusions can be made determinately compelling at a given time (and hopefully, at any other time). Within the framework of a quest for evidence, those problems are chosen and those methods are developed which lead to resolutions maximally determined by the methods used as well as the evidence obtained.

The view that there are meaningless problems is both a part of and a consequence of the scientific perspective. In science, problems may be formulated which are irrelevant because they are unresolvable. Surely there *may* (in some sense) be a Cartesian demon deceiving me, an invisible gremlin flying around inside my watch, or even extraordinary and divinely caused events. But they are not open to evidential confirmation. No determinate method of ascertaining their properties can be found. Science rests upon this kind of distinction. If we declare certain concepts to be cognitively meaningless, we adopt a particular attitude toward investigations in which they might be thought to play a role. Certain disputes are trivial, insignificant, or meaningless *from the point of view of science*. A statement must lead to, entail, or imply, in conjunction with other similar statements, an observable state of affairs in order to be scientifically relevant. Otherwise problems will arise which are hopelessly unresolvable. Science seeks to maximize the determinacy and compulsion of its conclusions.

It is important to mention an obvious objection to the position I am developing – that determinate resolvability is not a property of science any more than of philosophy: all scientific statements are corrigible. But resolvability is not the same as verifiability. The mere fact that a statement is corrigible does not permit us to conclude that an alternative can be found which is equally satisfactory. Simplicity and fruitfulness are nonlogical criteria which are vaguely yet efficaciously implicit within scientific investigations to make a single hypothesis acceptable. In the law of gravitation, $F = kmM/r^2$ is not empirically distinguishable from $F = kmM/r^{2.000000000000001}$. But surely no scientist would hold the latter to be true on evidence that confirmed the former. It is the making of such implicit choices that produces what I call "resolvability." Resolvability is a functioning *ideal* of science whether or not a given problem can be settled. We expect that it will eventually be resolved. This ideal is not appropriate in other fields. We do not expect nor even desire that every educated man will, one thousand years hence, have the same taste and understanding of works of art.

It is quite true that we would like also to declare some problems or issues irrelevant to philosophy as well as science. Yet if we adopt too strict a criterion of resolvability for philosophy, it will reduce to science. The doctrine that the logic of science is all of philosophy adopts such a point of view. But there is another alternative — that of accepting the relevance of problems that are not strictly resolvable by empirical evidence as they are formulated, but which are settleable in a larger sense of providing the greatest and most comprehensive understanding of all existence, rather than of the narrow domain that science to be strictly compelling must deal with. The breadth of such issues, and the wealth of relevant considerations, makes divergence of method and conclusion unavoidable. In other words, we may preserve the classical distinction between giving evidence and giving reasons for holding a belief. Philosophy searches for adequate reasons, resting on a more comprehensive ground than evidence or faith. There is no question that philosophy need refrain from asking, but it will not nevertheless accept every answer as adequate. There *may* be a gremlin in my watch, but such a view promotes a radical dichotomy in my understanding of things. I may be totally deceived, but I cannot both accept such a view and make rational claims, even about my deception. I must, to philosophize, order and constitute my world as coherently and systematically as possible. A puzzle here is spurious when it promotes sheer incoherence, and violates any possibility of systematic understanding. But coherence is a concept that cannot be given a precise meaning; philosophy of necessity is multifarious and partly indeterminate in its analysis. Philosophy rests on an ideal of adequate belief based on systematic and coherent reasons, while science rests on an ideal of understanding which is based on full and sufficient evidence. Such evidence depends upon the prior existence of accepted conditions in the framework of scientific investigation.

Another way of putting this is that to be a scientist a man must accept the methods and conclusions of the scientific process, and it is this prior acceptance that renders the problems he investigates determinately resolvable. But in philosophy, there is no such obligation and no such resolvability. A particular school of philosophy may make a number of antecedent choices to render some philosophic problems resolvable. Here philosophy comes much closer to science. But it is always appropriate to repudiate any particular point of view in philosophy, and to set out anew, with a new set of basic categories that creates a new world perspective. Such options provide the unavoidable indeterminacy of the philosophic perspective. Like art, philosophy makes a virtue of multiplicity.

Science, however, is uniquely compelling in the sense that there is a basic and expanding core of scientific material which must be accepted if one is to be a scientist. The great body of assertions in science (excepting only those at its frontiers) are unquestionable to anyone who adopts the scientific perspective seriously. In this sense science is uniquely compelling. Virtually none of the judgments of philosophy or art can be thought of as so secure that their repudiation amounts to a departure from the perspective.

The compulsion of evidence is in part produced by the logical rigor of science, for only where methods are rigorous and logical tests strict can evidence be found which is decisive in testing proposed hypotheses. Karl Popper claims that in fact science cannot verify its hypotheses, but can only falsify them.[13] The germ of truth involved here is, I believe, essential to the understanding of science. Science, unlike art, seeks rigorous negation as well as affirmation. Evidence is always pursued in science to *disconfirm* a current theory, for then the theory may be modified to conform to that new evidence. If evidence were found which only confirmed the theory, there would always be the possibility that we had not looked widely enough. No limits to the theory would then have been found, and the determination of such limits is central to scientific investigation.

Let me not be thought to be saying that science only negates and does not affirm. That would be absurd. Rather, affirmation in science is partly derivative: we affirm a theory when it has not been negated by rigorous testing. The goal of scientific research is to find experiments which negate a currently important theory, exposing that theory to critical analysis and modification. Lee and Yang won the Nobel Prize

[13] K. Popper: *op. cit.*

for *disproving* an accepted principle (of parity). The disproof is far more illuminating than evidence which only supports an already well-supported theory. The Michelson-Morley experiment was revolutionary in science because it shattered well-confirmed principles and laws. The strict logical rigor of science permits the affirmation of what has not been negated. That is how science uses the law of excluded middle. By using inductive methods in conjunction with that logical principle, science can assure itself of what has been disproved, and affirm its negation.

In art, however, every work affirms itself, without negating anything else. That is the source of the diversity of artistic validity. It is also a mark of the absence of logical rigor. An artist proposes a new art form which in no way replaces or negates any other form of art. This may well be one of the fundamental differences between the perspectives of science and art. That of science is unique because it is conceived to render affirmation in terms of negation. Art affirms without negating.

The compulsion of evidence stems from the hypothetical character of science, in its acceptance of assumptions that create problems resolvable in some definite or compelling sense. Hume's argument that no general law can be logically confirmed by any set of empirical evidence implicitly reveals his recognition that the evidential relationship is never absolute, nor can it be removed from the particular investigations in which it is defined by factors taken for granted. From a static point of view we seem to assume some principle of the uniformity of nature, which is vague and indefensible precisely because it means something different in every investigation. What is taken to be uniform in nature stems from the conditions taken for granted in the conception of the problem under consideration. We may conclude, for example, that all bodies fall with constant acceleration near the surface of the earth (vague as this formulation is) on the basis of a single experiment in which we observe that two unequally weighted balls fall equal distances in equal time, provided we make the appropriate prior assumptions of physical facts and mathematical relations. What an individual who studies experimental method must learn is why a particular experiment, though not a logical proof of a proposed hypothesis, is nevertheless confirmation of it. He learns to take the proper elements for granted. There is no general evidential relationship that holds between hypotheses and evidence confirming them; but in a situation governed by certain assumptions, only one conclusion is legitimate. Later, in other investigations, the original assumptions may prove inadequate in some respects.

This is why a scientific hypothesis which purported to claim that one of Newton's laws of motion was completely false concerning quite ordinary phenomena would be totally without value in a scientific context. If a man built a machine which violated the law of conservation of momentum, it would be no more than a curiosity until scientific analysis had reached the point where enough theoretical power had been generated by other laws and observations to call in question that principle or to preserve it by the introduction of hitherto unknown forms of matter or energy. Otherwise, everything said about mechanical phenomena in a scientific context simply assumes the validity of the third law of motion. There is, for this reason, a natural tendency to reject such a machine as impossible even when confronted with it – there must be some hidden trick – by virtue of the engrained habit of taking certain principles for granted. Even if the machine were a fact, it could scientifically do no more than pose a serious problem. Scientists would still be unable to make anything of it until resolvable problems could be raised concerning it – until hitherto unknown evidential relationships had been revealed which when directed into new investigations gave insight into the principles governing such a machine. They would otherwise not have sufficient conceptual strength to speak intelligibly about it. At best, it could only be described at the level of ordinary language, giving its most superficial and scientifically least valuable aspects. Although ordinary experience constitutes a relatively well-ordered and coherent system within which anything can at least be described, such a description may often be quite misleading and erroneous, as in illusory phenomena. Scientific discourse, if it is to be significant, is and must be theoretical in character. Sheer "facts" may raise problems, but satisfaction is achieved only when fairly comprehensive theoretical systematization is realized. Anomalies may exist for long periods of time alongside theoretical systems which they seem to contradict; they become significant only when new systems are developed within which they can be included, conceptually and evidentially. The scientific ideal of a comprehensive theoretical system within which everything has a place is not a blind faith, but the very fabric of science. Sheer factuality is chaotic and random, disorganized and unscientific, until related in a theoretical system that provides pathways of evidential relationships. The very nature of significant problems in science is defined by such systematization. Evidence exists only in theoretical structures of some degree of breadth which determine what will count as evidence.

Theories are not static. They change within the scientific process. Science is fundamentally developmental in character, for evidential relations develop in time and are transformed through inquiry. The view that science reaches the bare truth of reality cannot comprehend the possibility that certain problems may be not only practically, but theoretically outside scientific concern at a particular time. This view maintains that all we need is to study the world and understand it. But when science is seen to be a cumulative process, it is clear that some valid conclusions may not be drawn until an advanced level of development has been reached. Only when certain "facts" are assumed can certain questions be asked and answered by the collection of appropriate evidence. What is most important is that the answering of some questions reveals new "facts" to be utilized in the investigation of other, more remote and detailed problems of scientific concern. This constant modification of conclusions limitedly reached is science; and its fruits are realized in organized theories which coherently portray the evidential relations among its various parts.

It would appear that science is so compelling as to produce virtual unanimity of belief in its conclusions. Yet a considered analysis of the issues reveals that the state of affairs may be reversed: the compulsion of science may arise from the ideal of unanimity. When we speak of evidence in science, we refer to a body of statements and observations which are already defined by the criteria in use to be identical in import to all who claim to use scientific methods. In order that the resolvable problems of science exist, they must be resolvable to all who accept the same methods, principles and goals.

Thus, science cannot permit great divergence in the scientific community. If every scientist took an alternative set of facts for granted, the formulation of publicly acceptable conclusions would be impossible. Science is a shareable perspective, and its development depends on a community within which it is pursued. It is pleasant to declare that scientists desire understanding of the world, but also quite empty. A scientific conclusion is valid in a particular situation because the conditions of the situation render it necessary. It becomes part of the scientific heritage only when it is accepted by the scientific community at large. Such acceptance is forthcoming only when the conditions of the situation in which it is reached are accepted by that entire community. The conditions of inquiry are not private or personal; unless they are taken for granted by all scientists, they cannot be effective. Scientific progress rests on fundamental agreement at the many different levels of conditionality in the scientific enterprise.

I am not denying here that a principle is accepted by scientists because it ought to be; I am not claiming that acceptability is defined by acceptance. I am merely claiming that the principles of science are acceptable only in terms of some other criteria *actually accepted* in scientific practice. It can only be argued that a given conclusion is valid and should be agreed to by all rational beings on the basis of principles constitutive of rationality. If these are justified also, their justification will be given in terms of other criteria presupposed. At some point the justifying criteria are accepted generally by all rational beings, or justification altogether fails.

Fortunately, there are both facts of ordinary common-sense experience and rather general logical principles which are taken for granted by everyone who professes to pursue scientific methods. But it seems quite obvious that they are presupposed only insofar as they constitute essential elements of the scientific perspective – essential because it would make no sense to provide evidence for beliefs unless such principles and facts were accepted. We reject the claim that all sense-information is delusive for such reasons – that inquiry takes for granted the validity of some observations. Otherwise there could be no science. Education in science to a great extent is the building of sophisticated instruments based on the vague but fundamental sense of ordinary rationality and common-sense evidence. We become aware of the prevailing characteristics of evidence. We learn specialized techniques of utilizing conditions of scientific investigation. Above all, we learn to use only those methods that provide evidence acceptable to everyone. Creativity in science must be carefully channelled and ordered. If an individual asks the wrong questions, or asks a question at the wrong time, he ceases to be engaged in scientific investigation.

The perspective of science depends on the concept of evidence; but evidence is meaningful only in theoretical contexts where appropriate presuppositions have been made. Only when such presuppositions are communal, and underlie a social procedure of validation, can scientific conclusions be compelling. To one man, a particular assertion may be compelling because of his unique disposition. He may have particular biases which produce such compulsion. He is, to this extent, unscientific, however correct in his insights. Insight alone is not evidentially relevant, and evidence depends on principles accepted by an entire community. Individual creativity and uniqueness of point of view are essential in science as in art, but they must be ordered in particular channels. Scientific validity is social, not individual.

Yet individual contribution cannot be eliminated from science. If everything were routine, discovery would be meaningless. Something must be unknown, and not generally accepted, for investigation to arise. If only routine methods are employed, risk is diminished, but so is individual capacity to question and discover. Science depends on a shared perspective within which conclusions are compelling and significant, but the community of attitude can be overdone. Often individual recalcitrance and obstinacy of viewpoint produce the deepest scientific insights. Galileo had to break with his entire intellectual tradition to make modern science possible.

A scientist must accept certain common conditions in order to be a member of the scientific community. Yet in order to make a significant contribution, he must ask a question that has never been asked before, in the context of those conditions which define the community. Scientific discovery rests on turning conditions into problems and transforming facts into conditions. This is a precarious and dangerous activity, for the risk is not only of failure, but of treachery. If we raise a novel problem, we must attack some common conceptions still part of the communal background of investigation. This is the monumental risk involved in attempting to understand our world.

It does not follow from the recognition of the fundamental sociality of the scientific process that scientific validity depends in any way on arbitrary communal assent, that scientific validity is to be measured by taking a poll. Everyone *can* be wrong. Scientific methods are devised to be independent of individual biases, to transcend individual differences, to overcome the appeal to the majority will. An individual scientist can pursue scientific validity independent of all the desires and influences around him. But only by utilizing criteria accessible and intelligible to the scientific community. That is, he as a scientist operates in a framework partially isolated from the rest of his social existence, which is dominated by accepted modes of practice. If, however, he proffers a novel hypothesis, attacks a well-attested and widely accepted generalization, questions what has become a condition of contemporary science, he runs the risk of ceasing to be a scientist altogether; he may no longer be able to communicate his insights to others. The burden is his, to produce and facilitate communication. The existence of resolvable and determinate questions rests on a community which collectively accepts the results of its procedures. Of course, any fact may be tested, even modified in time. But only within another context of what is collectively intelligible and persuasive. The invention of a new system, conceptually novel and methodically unique, would be incommunicable inventive caprice.

F. Theoretical Validity

The view that what constitutes evidence for a theory is inevitably rooted in the historical and theoretical development of science suggests that no definite statements can be made concerning the grounds of validity of a theory on the basis of evidence. The criteria vary with the theory under consideration and the time of its evaluation. This position is correct, if endowed with the proper qualifications. It is often maintained that the function of a theory is to link together observed events, that the goal of science is "to construct a single network of true general principles relating events to one another." [14] But it can be shown that such linkages are not directly comparable; the events related are different in different theories. Evidence is always partly defined by the theories which are under test by it. The deductive model's view that a theory is adequate if it provides a means for relating all observed events is notably insufficient if it holds that observations are theory-neutral. The presuppositions, historical compulsions, and conceptual framework within which the observations are taken at least partially determine them. No theory is ever under test by all observations, and testing is not simply a matter of comparing a theoretical system with the world through experimentation.

The difficulty which arises here is that an answer to the question "which theory is more valid under the available evidence?" becomes vague if not incoherent. If the criteria used vary with each theory under consideration, or at least with each historical development, what constitutes a more valid theory?

It may be noted that no two cotemporal theories held to be valid in science have been so completely different as to possess no problems or phenomena of decisive importance in common. Fundamental to Ptolemaic astronomy, Newtonian cosmology, and the theory of relativity are the phenomena of planetary motion. The radical conception of Galileo that terrestrial and heavenly motion were in principle of the same kind marked a major departure from medieval cosmology, but by no means to such an extent as to rule out observed positions in the

[14] I. Scheffler: *op. cit.*, p. 55.

sky. The development of new instruments – notably the telescope – transformed observations of the night sky. But the primary assumption, that naked-eye observation and ocular observation revealed the "same objects," has never seriously been questioned. It is often said that the replacement of one theory by another takes place only when the old can remain embedded in the new in some important sense.

Clearly classical dynamics is not "true" for some range of phenomena narrower than had originally been supposed. It is simply not "true" for any range of phenomena at all. It is, however, unquestionable that certain definite problems (of the motion of bodies in a single frame of reference) are resolvable to complete satisfaction by the older theory. There is no possibility that Newtonian dynamics is unrestrictedly valid; but there are problems which Newtonian dynamics can solve. It is thus valid in the sense that problems can be satisfactorily resolved by use of it. Such problems have not been abolished with the advent of relativity theory. Moreover, the newer theory reveals a definite mathematical procedure whereby the older theory can be derived from the newer under assumptions which pertain to the problems mentioned. A "better" theory replaces an older one only when the success of the older can be preserved by such embedding.

What then of the problems dismissed by the newer theory as irrelevant, unscientific, or misconceived? Problems concerning absolute space, time, and mass are eliminated by the theory of relativity. Galilean dynamics dismissed the teleology of older views and at the same time repudiated the comprehensiveness enjoyed by Aristotelian cosmology. Nevertheless, in both cases, fundamental problems carried over from the older theory to the new, some of which the older theory could solve, some of which it could not. In the historical context of the time, there was a definite improvement in the number of problems solved and the precision of their solutions. Where two cotemporal theories are compared it may perhaps be said that so far as a fundamental set of problems is concerned, one theory is clearly better than the other. A theory is more valid than another: (1) in that it more satisfactorily solves problems common to both; and (2) in its fruitfulness for defining and resolving hosts of new problems on the basis of evidence. Statistical thermodynamics replaced classic thermodynamic views only by preserving the older views as deducible from the statistical hypothesis, and at the same time leading toward both a new conception of primary elements in chemistry and the possibility of observing molecules.

But when we step back from the moment to compare theoretical world-views of different historical periods, the above picture of the evaluation of theories fails. How compare Aristotelian cosmology with that of the theory of relativity? Their goals, criteria, principles, and critical phenomena are worlds apart. The norms which grounded Aristotelian explanation are set aside in modern physics. The latter is immensely more precise over a much smaller range of phenomena. If it is "better," it is so only according to the criteria of scientific adequacy that are now accepted, and which would not have been acceptable eight centuries ago. We may dismiss Aristotelian cosmology as bad science, but only in terms of a conception of science developed historically through the surmounting of older theories by newer ones, under the immediate pressure of critical problems.

Science is irrevocably historic in its methods, assumptions, and goals viewed over the breadth of its development in time. But it does not follow from this that the replacement of one theory by another is an arbitrary or willful act, nor even that it is at the whim of the scientific community. Rather, each such replacement is a definite and justifiable act in terms of the nature of science at that particular time and the problems at issue then. Some embedding is even carried throughout history. But the natural development of science, filled as it must be with preconceived notions of significant problems and phenomena, makes it impossible ever to find but apparently arbitrary grounds for comparing two different world views. In this sense, what science is at any time cannot but be a consequence of its historic development. Science today is not the science of Newton or Galileo, and certainly not of Archimedes.

G. Objectivity

It is time to consider the most well-known of the properties of science, even more important to the average man than its empiricism – its objectivity. To the layman this may be the distinctive property of the scientific perspective. The methods and conclusions of science are objective; art and theology are subjective, if not personal and private. In a world which is essentially realistic in its fundamental attitudes, this objectivity marks the superiority of science. Where subjectivity rules, truth is abandoned.

What is the objectivity that science possesses, and is it superior to what is called "subjectivity" in its cognitive attributes? May we identify objectivity and reality, so that the objectivity of science marks its close relationship to the real qualities of things? Such an identity neglects the critical dimension of Kant's metaphysics: the identity may be considered significant only if both terms can be given clear and definite meaning. Unfortunately, the term "reality" is without independent meaning. In fact, a proper analysis of the identity shows that the objectivity of scientific theories is the fundamental concept, and that the reality identified with it has no significant meaning of its own. Put another way, there is no test in experience as to whether a well-confirmed scientific theory corresponds to reality in any other sense than that it is indeed well-confirmed. Nothing can prevent us from equating objectivity and reality, but when we do so, we once again *define* knowledge as science. This does not provide great clarity or insight.

Kant himself simply dismissed the metaphysical problems involved here, and found the ultimate grounds of science to lie in the apodeictic certainty of scientific laws. If scientific laws were necessarily true, then that solution would be a satisfactory one. Unfortunately, they are not. Kant also identified the objectivity of science with the *universality* of scientific conclusions. The difference between the subjective and objective rests upon the public validity of the latter. What is subjective is personal, perhaps (though not necessarily) private. What is objective is universally compelling, binding in effect on everyone.

Such a view has great strength if the universal compulsion of science rests upon fundamental principles of human reason, if the universality of scientific conclusions is a consequence of their nature. So long as there is something necessary to human thought, it may be considered certain and universally compelling, and the three together may be identified with objectivity. However, if there is nothing permanently necessary in human thought, we are left with a much weaker and less plausible equation – of that which is universally accepted and that which is objective. We cannot even speak here of universal compulsion, for only necessity compels.

I have given an account of both the particular mode of compulsion to be found in science and the universality of scientific conclusions. Science is indeed uniquely compelling, both psychologically *and* perspectivally. However, psychological compulsion provides no grounds for the superiority of science. It is at best a causal factor in the power and sweep of science in modern intellectual life, and quite irrelevant to the claims of science to objectivity. Perspectival compulsion is much more to the point, and must be considered carefully.

The universal appeal of scientific methods should not, however, be passed over lightly. Where a work of art makes no claims at all to appeal to everyone, even to all aesthetically and trained observers, science seems to demand acceptance by all who understand it. This is due to its particular mode of compulsion, and will be considered in a moment. However, it must be repeated that the powerful compulsions in science never actually lead *every* first-rate scientist to acceptance of all the conclusions of science at any time. Universality is not a property of scientific theories, but a goal toward which science aims. Science pursues universality; it does not claim it. At best it can only claim that everyone *should* accept its conclusions, which is another matter entirely.

It is impossible to defend the claim that everyone should accept any *particular* conclusion of science. Let us not forget the fundamental corrigibility of all scientific statements, the tentative nature of scientific theories, the likelihood that any particular scientific formulation will be abandoned as science develops. The classical theory of motion was as compelling a theoretical system as could be imagined on the basis of evidence, yet it was abandoned on excellent grounds. Not only are few scientific theories actually compelling to *everyone*, but it is indefensible to claim that they *ought to be*. Perhaps science generates universality, but tempered profoundly with the realization that nothing *in particular* in science can be deemed final and universally binding, or we stop the process then and there.

Moreover, it is quite implausible that we should feel compelled to accept scientific conclusions *because* they are accepted universally. The reverse is in fact true: we may assert that *some* essential core of scientific conclusions ought to be accepted by all scientists, which we cannot quite determine exactly, because of the particular and definite compulsions of the scientific perspective. We are led again back to the compulsion of science as the ground for its objectivity.

Unfortunately, the perspectival compulsion of science offers no grounds for a principled distinction between subjectivity and objectivity. The scientific perspective has its unique mode and degree of compulsion. It minimizes divergence and disagreement and aims at the maximization of determinacy. But not only does it fail to produce unanimity (nor could it), it in no way can rule out all alternatives. Every perspective has its own modes of compulsion. Not every painting is a great work of art; not every novel is a masterpiece; not every act can be judged good and right.

The distinction between the subjective and objective suggests that all warrant belongs to the latter, while what is subjective is warrantless, devoid of justification. Such a view is quite insupportable. All methodic and intentional perspectives have modes of compulsion, if not so universal and determinate as that of science. Perhaps art seeks a great variety of aesthetic forms, but within that context there are definite modes of compulsion. Few works of art are of profound significance. If art is subjective, it is not arbitrary or completely idiosyncratic. A trained musician in no sense presents an arbitrary reaction to music: he presents a careful analysis that reveals the compulsions inherent in the perspective – appreciative or technical. If science is objective, and we identify objectivity with compulsion, then science is at best *more* objective than art, but not wholly so, and art is less objective than science, but not completely subjective. But this distinction is so artificial that we must sometimes labor to preserve the claim that the warrant achieved by an artist or critic is less than that by a scientist. For example, the view that the universe is expanding is less defensible than is the fact that *Hamlet* is a masterpiece. The identification of objectivity with compulsion breaks down once it is recognized that there are many modes and degrees of compulsion. The distinction between the subjective and objective in methodic disciplines is not a useful one.

The only view that can be defended is that science seeks a particular *type* of compulsion – that which minimizes, but does not eliminate, divergence. It pursues a mode of validity that determinately grounds particular claims. The ideal of science is the presentation of a systematic conception of the world which has no valid competitors. But this world view eliminates its competitors only within the perspective of science. Outside that perspective (but not outside the domain of objectivity), alternatives remain that cannot be eliminated. Nothing can compel a man to be scientific. He will be scientific only if he seeks the kind of compulsion science allows. Other modes of compulsion, with their types of validity, always remain.

If we equate objectivity with the perspectival compulsion of science, we may say that science achieves objectivity by methodically eliminating divergence of outlook and covalid alternatives. Properly speaking, science is not superior *because* it is objective; it is objective because it adheres to an ideal of determinacy and resolvability. Objectivity in itself is not particularly good, but the decisiveness and security that come with maximal compulsion are. Objectivity is a property of those methods which achieve the greatest determinacy in human experience. It does not follow that this determinacy is either the greatest good in human experience, nor even a very important one.

The objectivity of science is sometimes contrasted with the normative qualities of art and religion. Science is held to be value-free – that is what "scientific objectivity" means: freedom from bias, from personal concern, from particular values. Putative sciences like psychoanalysis, history, and sociology are often criticized for being fundamentally normative, and therefore unscientific. Here universality may be brought back into the discussion, not as an essential property of science, but as a goal which is impossible so long as personal values are part of the scientific perspective.

There are two senses in which one may speak of science being value-free: (1) it is value-free in that no normative judgments of *any* sort have a place therein; (2) it is free from the impact of moral, social, or personal values, which are sometimes called "prejudices" or "biases." Completely different issues arise in these two cases, and they must be distinguished.

The discussion given throughout most of this book suggests that (1) is quite false. The perspective of science is defined by conceptions of what is good and bad in science. Within the perspective, objectivity, compulsion, and warrant are definite values. Scientists quite clearly make normative judgments about the worth of scientific methods and their results.

More interesting than this, however, is the view that in addition to the defining normative conditions of the scientific perspective, there exist values essential to the pursuit of scientific investigation. It is plausible to claim that science is objective at least in the sense that the scientific investigator must be free from biases in his investigations, though he may be heavily prejudiced in favor of the perspective of science. Polanyi, however, makes very clear that inherent in all scientific inquiry is a personal commitment, not merely to science itself, but to a rational sense of order in the universe. There is nothing mechanical and routine about scientific investigation, but rather the active engagement of a person's intellect with the world. Precisely because science depends upon the personal mind questing after the truth of things, it cannot be thought of as a routine act but must be recognized to be personal, individual, and normative. Great minds leap the bounds of ordinary scientific theory. "The original mind takes a decision on grounds which are insufficient to minds lacking similar powers of creative judgment. The active scientific investigator stakes bit by bit his whole professional life on a series of such decisions and this day-to-day gamble represents his most responsible activity."[15] Science is a revolutionary movement from theory to theory, and depends upon the power of the individual mind.

Yet the fiduciary aspect of scientific discovery emphasized by Polanyi does not make science subjective, for there remains objectivity in the sense of "affirmation of personal conviction with universal intent."[16] By accepting almost a Kantian solution to the problem of scientific objectivity, Polanyi affirms the normative foundation of science amidst its objectivity. My own view, however, based on the analysis of objectivity given here, is that this is quite unnecessary. The personal component of scientific discovery is and must be followed by the development of procedures that render the novel theory determinately compelling or valid. If no means exist to validate a given theory, then however powerfully an individual genius defends it, it will remain outside the perspective of science. Psychoanalysis inhabits the fringes of psychological *science*, however central it is to our conception of man, for it cannot be grounded in compelling evidence.

The conditions of scientific validation, which at any particular time define the methods of science, must be transcended by men of genius who revolutionize scientific thinking. Their risk – a risk of competence – is a normative one; they wish to transform the values at the heart of the perspective of science which surround the conception of evidence. When Galileo proposed that the heavens and the earth were to be thought of as one, not two systems of evidence, he revolutionized the entire domain of science. Such a transformation is never a purely empirical matter, to which evidence is sufficient. The very nature of evidence is at stake. A scientific revolution is founded on rational ideals of evidence that are fundamentally normative. To Galileo, a single system of explanation was *better*.

There remains for consideration the question of normative judgments external to science, which intrude upon the scientific perspective. Is not science objective in the sense that the moral and aesthetic values of individual scientists have no place in scientific investigations? And if so, how are we to cope with the fact that many putative sciences are laden with moral and aesthetic values?

The conception of science as the perspective which seeks maximal

---

[15] M. Polanyi: *op. cit.*, pp. 309–10.
[16] *Ibid.*, p. 324.

compulsion and minimal divergence of judgment suggests, it seems to me, that in almost all cases where values arise, science must adopt procedures which eliminate them. In almost all cases of moral judgment, some degree of difference of opinion is legitimate and expected. Ethics as a perspective is by no means as determinate and univocally compelling as science; when the perspectives are mixed, that of science becomes indeterminate to the degree that ethical considerations arise.

I use the word "almost" in the above paragraph, however, to emphasize that it is a rather indefensible assumption that in *every* moral situation, some difference of opinion is to be found. If is it true that all human beings share a community of moral judgment, by virtue of a common biological and social heritage, then these common moral elements can enter the domain of science without in any way affecting its compulsive structure. This is the justification for the view that the clarification of moral standards, and the scientific analysis of the conditions of moral judgment, can lead to moral progress – by finding a common heritage of moral commitment which all human beings *as* human beings share. Once this common moral domain is set forth, then it may well be relied upon, as a genuine property of things, and can play a role in scientific investigations as well.

But only undue optimism can lead us to suppose that all or even many moral issues which concern mankind are rooted in such a community of aspect. Wherever controversial values arise, even in principle, science is undermined, for its compulsion is weakened. If a particular domain of human investigation is essentially value-laden, then it cannot be scientific.

I hope it is clear, therefore, that such scientific objectivity is not only a great value, providing success and certainty, but a terrible liability. For the most critical and significant of human problems are heavily value-laden, and cannot be stripped of their normative character without trivializing them. This is not a minor matter, for the fundamental importance of the perspective of science is at stake here. Science maximizes determinacy by setting aside problems which are unresolvable. These problems are not eliminated; they do not vanish. They are simply ignored by science. Other domains of investigation must then preempt them; and it is these domains that become of critical importance in human life.

A number of examples might be useful here, chosen from a number of different disciplines. First, and perhaps most well-known, is the problem of determining the goals of various types of therapy. In particular,

I have in mind types of psychotherapy. These differ from physical therapies in that there is a definite and determinate set of conditions for physical well-being that all human beings do in fact agree on. This is a case where the common set of norms do indeed allow scientific research on medical problems, though *in principle* all sorts of difficulties could be raised. Is a man healthy if he is twenty pounds overweight, if he is pale, flabby, and so on? By setting these aside, and investigating only clear cases of illness, medicine can utilize the procedures of science. But mental health is not so easily determinable or defensible. Abnormality is often a virtue, especially in conforming and rigid societies. There can be no science of psychotherapy.

Yet surely it does not follow from this that no study of types, goals, and methods of psychotherapy is possible. It cannot be a science. Rather, it must be a rich mixture of philosophical analysis, moral judgment, and empirical investigation. Like most of the social sciences, it must let its fundamental concepts remain value-laden, and work within a context where values are taken seriously, and treated explicitly. Aristotle's ethics is irrelevant to quantum mechanics; but it is vital to any understanding of mental health. Here is a domain of understanding that is unscientific, which is a vital part of human life, yet which can never eliminate diversity. It is precisely because every attempt at a theory of psychotherapy is followed by critical disclaimers of a normative sort that we continually adjust our understanding of man and our goals for him.

It is unfortunate that the reputation of science has grown so inflated as to make us think that only experts of scientific authority are entitled to make technical decisions on political matters. It is held that only "experts" may speak out on issues of foreign policy, as if technical knowledge of the countries involved were all that mattered. In fact, however, questions of foreign policy are also a mixture of normative, technical, and scientific components, a mixture that cannot be separated into its elements. Political decisions can only be made on the basis of a fusion of moral, philosophical, and empirical factors. Experts in any one or another particular area have no more qualifications on which to base decisions than a reasonably well-read and morally sophisticated layman. In fact, if they are professionally amoral, as scientists, then they are far less qualified.

In the psychology of learning, ordinary conceptions of intelligence, particularly its normative properties, seem to dominate the field. It is better to be intelligent than not: everyone takes that for granted.

Unfortunately, precisely because everyone agrees about this, but fails to agree on just what is in fact worthwhile, psychology fails to deal with the normative issues raised. Many psychologists escape the moral problems involved by *defining* intelligence as "that which is measured by test W." Obviously, this is fruitless from the point of view of any significant problem in human life that is related to intelligence. Yet it may be necessary to scientific investigation. If it is, then it is quite clear that some other study, thoroughly value-laden and unscientific, must integrate the value properties of intelligence in human life with the psychological facts of human learning.

Finally, to pick a very different example, consider the question of the origin of the universe. Science can investigate the conditions from which a later stage of the universe or any part arose; but that is not what a theologian has in mind. What is sought is a metaphysical, rational, and moral conception of the history of things, so that one space-time epoch can be grasped in a total vision of its multiple aspects. Science cannot provide this. It may well be that poetry, philosophy, or theology can. Are we compelled to accept a poetic vision, a philosophic system, a theological world-view? Not as we are compelled to accept a scientific law. But there exist all sorts of compulsions, and some of these may be decisive in one sense or another. Even if there is no determinate system possible, is not the quest for one valuable?

In concluding this topic, it may be worth asking how science achieves whatever objectivity it does achieve – or, put in terms of the categories appropriate to this essay, how does science avoid particular biases that would weaken its mode of compulsion and create divergence of point of view?

There are two major factors at work. On the one hand, by working within a community of effort, where as much as can possibly be shared in common is in fact shared, the particular biases of any particular individual are minimized if not eliminated. Here logical analysis and methodology (literally, the *study* of method) are fundamental: to define the terms and conditions of scientific investigation so that particular judgments of individual scientists are minimized, so that scientific validity becomes as routine as possible. By adopting systematic presuppositions which define the nature of evidence and the grounds of validity for scientific investigation at any particular stage, different scientists generate as uniform and determinate a mode of compulsion as possible. This same compulsion represents a condition of investigation within which the particularity of individuals is minimized, and their community of interests and methods maximized.

Divergence arises in human experience from genuine differences among human individuals. Idiosyncracy and particularity are as clear traits of human experience as are community and similarity. The fact that, however scientific methods are, they contain an irremovable residue of individuality is captured by Polanyi when he emphasizes the unavoidable contributions of individuals of genius in the process of science. On the other hand, it would be a mistake to ignore the fact that science utilizes methods which render the individual components within scientific investigations minimal and often quite irrelevant. It does so by working with its history in such a fashion that the grounds of scientific validity compel assent. Science is then unavoidably historical, a trait which is the source of its hypothetical nature. It is also the source of evidential compulsion. The objectivity of science in this sense is a result of methods which produce an evidential community. All scientists do indeed accept the same events and observations as evidence. They must do so if they are to be scientists. And they are thereby rendered objective.

John Dewey has a wonderful passage which captures the situation here, though it pertains only implicitly to science: "a human individual is distinctive opacity of bias and preference conjoined with plasticity and permeability of needs and likings.... The individual that finds a gap between its distinctive bias and the operations of the things through which alone its needs can be satisfied ... either surrenders, conforms, and for the sake of peace becomes a parasitical subordinate, indulges in egotistical solitude; or its activities set out to remake conditions in accord with desire. In the latter process, intelligence is born."[17] Science is the perspective which depends upon methods which minimize the function of opacity of bias (though they cannot eliminate it entirely), and maximize the role of what is shared. To Dewey, such a perspective is the only intelligent one. I have suggested that such optimism is an error, that it is intelligent to recognize where individuality has a place, and to respect it. Dewey did not recognize that other methods than the scientific could exist and could be intelligently pursued.

The objectivity science achieves by avoiding ordinary values utilizes a special device – that of abstracting from particular concerns to the point where individuality is of minimal importance. From a practical point of view, only particular things and people matter. But particularity is

[17] J. Dewey: *Experience and Nature*, pp. 242–245.

of no scientific relevance. In seeking general principles, the individual scientist withdraws from the immediate environment and works with abstractions that have far less emotional and moral impact.

In the physical sciences, personal involvement is far less obvious and striking. No obvious violations of ordinary attitudes is necessary to a physicist or chemist. Yet a consideration of theologies which have committed themselves to particular cosmological pictures reveals that the physicist's objectivity is an artificial device. Even in physics, the scientist must learn to ask the proper questions, and to avoid problems that create divergence of point of view. He learns which particular properties of things are amenable to objective study – meaning nothing more than which problems are open to determinate resolution.

In the social sciences, however, moral considerations enter so easily and naturally that the scientist must violate the most elementary forms of human feeling and moral concern. The research psychologist cannot view his subjects as his friends for his emotional reactions would predominate. He identifies them as members of humanity, as abstract entities, as tokens of a general type. His research may lead him to conclusions that he personally abhors – but he avoids a confrontation of this by denying the relevance of the research to his personal life and feelings.

Put another way, science is the *use* of particular things to reach general principles. But as Kant pointed out, it is immoral to use human beings in any way at all. What science must do, then, is explicitly to set out in an amoral (if not immoral) framework. And it does so by ignoring the particular human beings concerned in its work. This objectivity, rooted in a repudiation of morality, is the foundation for scientific research on weapons of war of the most atrocious sort. It is one of the prices to be paid for scientific investigation. It is well-known that many scientific experiments are cruel and appear to be malicious, particularly in psychology and medicine. It is not obvious that this is always due to intentional cruelty, but is probably a further consequence of scientific objectivity. The latter can be too great; we may pay a fearful price for scientific progress.

H. The Rational Life

There are countless ways in which men experience their world, and science is but one of them, though with philosophy, morals, religion, art, and statesmanship it constitutes one of the great perspectives. Nevertheless, it is but one of many, and must play a role congenial and

harmonious with other modes of experience. Men may search for belief and for evidence for their beliefs; but belief alone will not content
them. They also search for significant ways of making things, forming
structures in paint or words, for comprehensive visions of life, for
ways of evaluating things and actions, for an ineffable consciousness of
belonging to some overarching order, for a coherent system of power
and social relations. None of them alone is capable of constituting the
totality of human experience, though some men have made one or
another of them the primary focus of their lives. Within any society,
they interact, comprising the culture of that society. Each of these
perspectives is a way of ordering things; the scientific perspective is
not alone able to provide order; different men prefer different orders or
different emphases.

Yet science is uniquely compelling and determinate. It rests upon an
ideal of determinate order. In Whitehead's words, science rests upon "the
instinctive faith that there is an Order of Nature which can be traced
in every detailed occurrence."[18] An artist can never entirely determine
a single mode of judgment, however he limits his scope of interest; a
philosopher may adopt alternative ways of analysing things to render
them meaningful and intelligible; men worship their gods in various
and diverse ways; ethical judgments possess an irremovable openness
even when made responsibly and with sensitivity; different forms of
social order preserve cultural divergences that are both necessary and
desirable. Only science encourages the claim that under certain conditions only one truth is possible; given a scientific community growing
out of an actual history, only one conclusion may be drawn from evidence acquired by rational and tested procedures; given a particular
event, only one explanation is legitimate. If we cannot realize such determinacy now, we expect to attain it sometime in the future. Such determinacy, however, rests entirely upon the antecedent acceptance of
the scientific perspective in many quite specific details. The determinacy
of science results from taking the perspective for granted. Nothing,
however, compels anyone to accept the scientific perspective. He may
simply repudiate that way of experiencing the world – and, of course,
be limited by it: but why may not such limitation be viewed as a good?

The faith perseveres that science is somehow the essence of the rational commitment, that reliable judgment can only come from the active
use of intelligence scientifically directed to the solution of problems,
that science is the only method for those who endeavor to understand,

18 A. N. Whitehead: *Science and the Modern World*, New York, 1925, p. 6.

to predict correctly, or to adjust to and use the things of their world. Science is not only compelling to those who have already accepted its perspective; it rationally compels all who wish understanding.

Problems open to determinate and precise solutions lie solely within the province of science. The methods of science are precisely those methods for providing resolvable problems with solutions. To this extent, insofar as adjustment is based on precise control and prediction, and insofar as prediction is conceived of definite and limited scope so as to be completely determinable by experiment and observation, only scientific methods and criteria are applicable. Moreover, as science develops, it preserves the solutions to problems once solved. It incorporates them in new theoretical formulations. In its own domain – of determinate and resolvable problems – science rules supreme. In prediction, we forecast a precisely determinable event; the determination of our expectations entails scientific inquiry. Insofar as we demand a single and decisive rule for adjustment, insofar as we search for a determinate path through the exigencies of life, science is our guide. If we demand from the world one way, and emphasize its determinacy, then we are compelled to the methods of science. In this sense, science is uniquely compelling.

But there are other meanings of "adjustment," "prediction," "understanding," and even "control." Art and religion are ways of responding to the powers of things, and making our way among the realities of life. A successful painting not only is a way of responding to an aspect of experience, but also orders and characterizes its future. The pervading mystique and the authoritative canons by which a man affirms his allegiance to the divine, structure and characterize his entire life. These are methods of adjustment, of the incorporation and transformation of experience. Events and facts are infused with determination by such perspectives. A great novel, filled with profound psychological insights, offers the possibility of complex and subtle anticipation of future experience. A moral treatise, close to the ethical needs of men, and responsive to their desires, provides moral understanding. Though not open to decisive test, it enters into future experience, creating it partly in its fashion, imposing upon it subtle moral qualities, transforming the ways in which men meet their moral concerns. A philosophic analysis, if illuminating, renders future experience more open to characterization and definition, clearer and richer at the same time. Surely this is a powerful means to greater control. Indeed, if "control" means the intentional direction of experience, then all intentional modes of human production are ways of controlling existence.

The great power of scientific control rests solely on the degree of determinacy it provides in its answers and conclusions, a determinacy which stems from the context chosen, the presuppositions accepted, the elements taken for granted. Science is unique in its affirmation of decidability as a value, and in its development of methods for extending its domain of application. But while all subject matters are open to scientific scrutiny, it is by no means obvious that all issues of human concern are determinable. Moral, political, religious, philosophic, and aesthetic issues may not be determinable beyond a certain point of multiplicity. That is why so many contemporary philosophers reject them as noncognitive. Yet they are quite susceptible to determination. We do pursue moral wisdom, religious experience, great works of art, and ideal forms of political order. We may abstain from participation in the great enterprises of experience, abdicating the moral imperatives of attempting to direct our own destinies. But such an abdication is also an abdication of human intelligence in some dimensions of experience. Science is not the only intentional mode of directed transformation of experience – it is only the most determinate one, devoted to those problems which are most resolvable. Perhaps the important problems – concerning the good life, the ideal society, the elements of the beautiful, the qualities of the divine, or the meaning of life – are necessarily somewhat indeterminate. They are either vaguely conceived, too close to our idiosyncrasies, or ambiguous with respect to evidence. But they are not closed to our understanding. Other modes of analysis than science seek their comprehension.

A man may simply refuse to enter the scientific perspective. He may not wish determinate answers, nor may he care to make precise predictions. There is no compulsion for him to relinquish his own point of view and become scientific unless he antecedently accepts the value of determinate and precise solutions to particular problems. There is no ground for declaring him irrational, unwise, or misguided. He may have come to a profound and insightful understanding of the indeterminable elements of human experience. He may be religiously inspired, possess a truly moral soul, and have found a deep source of meaning in life. He may produce finely-spun and lovely works of art, and be deeply aware of the peculiar ways of his fellow-men. It is true that he cannot predict the motions of the stars accurately. He lacks any but the most rudimentary knowledge of science. But if he possesses the rest of the world, does he lack for much? There is no compulsion on men to be scientific unless the domain of the determinate is of significant value to them. They must accept the value and purpose of science in order to be compelled by its conclusions.

Psychologically and socially, however, science has an undeniable compulsion, and to many who pursue other modes of comprehension. Within the educated culture that one demands if the great options are to exist as viable choices, science is the most secure, the most successful, and the most complete. This is partly because of its perspectival compulsion: relatively speaking, science is resolute and determinate, and provides secure and compelling conclusions. Amidst the precarious and ambiguous choices that permeate human experience, science is a haven of security. Its language is less ambiguous; its problems are not open to eternal doubt; what is unresolvable is dismissed from its consideration. Philosophy and art become pleasant but unimportant diversions; and the richness of human experience reduces to a domain of resolutions and expectations. Since the other perspectives of human experience do not vanish, they lurk as ghosts in the background, not open to methodic production, not ways of knowing.

Among those who have chosen to pursue the great human perspectives rather than narrow personal success, the spectacular powers of scientific methods have proved to be overwhelming in their compulsion. Science is overwhelming in its ability to provide success and security at the same time. We find that scientific standards enter domains far removed from their origin. Philosophy and history are often thought to be governed by criteria of resolvability that are quite alien to them, only because such criteria have proved so successful in the sciences. A measure of the value of an enterprise becomes its conformity to the scientific ideal. Only resolvable problems become legitimate, even in philosophy and art. All other human enterprises are relegated to the subjective. Only scientific criteria are worthy of the name, only scientific methods are actually methods.

The psychological compulsion of the perspective of science has led many humane men to the position that science is destructive of important human values, that it tends to render the sensitive values of life but a series of mechanical movements. Such a view takes for granted that science is an engulfing perspective, stifling all other modes of experience. Perhaps this appears to be so; but in fact, science cannot be destructive to a single value which is part of another perspective. There are countless ways of experiencing the world, and an indefinite number of human perspectives. They cannot be legislated out of existence. Their legitimacy is not dependent on a community which accepts them. Only when a poet claims to reveal the laws of physical nature or psychological facts about men, or a philosopher legislates the cosmological structures of the stars – intruding into the determinate domain of science – are they open to scientific counterattack. Art, science, philosophy, and religion are separate and distinct human perspectives. The richness of experience is revealed in their distinctness and integrity. All of them operate within accepted conditions that define them. Nothing in human existence is so absolute as to comprehend all other perspectives.

To view men as capable of diverse experience, to emphasize their multiple potentialities rather than an organized perspective within which everything has a place, is to adopt a pluralistic conception of human experience. An indefinite number of perspectives enrich human life without organization into a single structure. It is possible to have diversity without conflict if the great perspectives are developed independently within their own spheres. We sometimes need precise and determinate guidelines – here science is necessary. But we may prefer greater insight, the systematic relation of disparate elements of existence in less precise ways. Here philosophy is more valuable. And if we seek insight, aesthetic delight, without logical analysis or "truth," we may well turn to some form of art. Human experience is rich enough to include them all, provided they not encroach on the others' domains.

Yet if the goal of philosophy is a coherent, well-organized system in which all experience has a part, then it becomes the duty of philosophy to seek the reintegration of the great human perspectives, to provide a coherent order to experience. Can we not hope to unify the shattered elements of life, so to organize our lives that science, art, philosophy, and religion all have a specific place? Can we not methodically determine an overriding perspective which we may consider the ideal form of human experience? Is this not the goal of human understanding?

Unfortunately, the usual method of integration is to choose one or another of the great perspectives as more rational, more valuable, or more worthwhile in some particular respect, and to view the others as incomplete versions of it or surrogates to it. Such a conception achieves unification by severe distortion of the possibilities of experience. If we are to integrate experience in a felicitous manner, we must be clear that diverse perspectives have their own integrity.

Subsequently, however, we may realize that human experience has its rhythms and its needs, and that in the rhythm of these needs may be found room for all the richness of possibility. The human organism constantly requires the solution to specific problems of life. It may move from these into the domain of science, with its determinacy and precision. But choice is as fundamental as need, and evaluations are called forth both within and external to science. On the other hand, very specific questions can be raised concerning types of evaluation. Science, morals, and art form an interrelated set of perspectives. The shift from one to another aids both in clarifying and broadening our understanding. We may continue, to the loose ends and incoherences that call for philosophic analysis, and the depth of commitment that approaches the religious attitude. Indeed, there is the possibility that as each of these perspectives is capable of universal applicability, so the rational life is precisely the extension of all of them in their indefinite interaction. The only coherence possible may be that provided by the application of as wide a range of perspectives as possible to every subject matter. Thus scientific knowledge about psychology and choice is relevant to religion and morality, while a given scientific theory may be judged aesthetically or ethically. What good is science? How true is *Hamlet*? Is not the realization of the breadth of possibility the only truly humane life?

INDEX

Adjustment, 149f.
Analytic Statements, 28f., 75, 117n
Anaxagoras, 9n, 113
Archimedes, 9, 137
Aristotelian Science, 11, 13, 136f.
Aristotle, 9n, 10, 13, 112, 144
Art, 2f., 11, 16, 17ff., 43, 53, 103ff., 119, 123ff., 128ff., 134, 137ff., 147ff.
Assertion, 91ff.; see also Truth
Authority, 8, 11, 15f.

Barzun, J., 125n
Becoming, 40; see also Process
Bohr, N., 70, 72
Brahe, T., 14
Bruno, G., 11f.

Carnap, R., 22ff.
Causality, 82ff.
Classical Physics, see Newtonian Science
Commitment, 3, 5, 17, 28, 141ff.
Compulsion of Science, 106f., Ch. VI passim
Conditionality, 73ff., 78ff., 100, 121, 133
Control, 149f.
Copenhagen School, 72, 118
Copernicus, 11ff.
Corrigibility, see Fallibility
Creativity, 133f.
Cumulation, 64ff., 88ff.

Data, 4; see also Facts
Democritus, 81, 111, 113
Determinacy, 76, 84f., 107, 113, 139ff., 148ff.
Determinism, 60f., 82ff.
Dewey, J., 4, Ch. III passim, 89, 104ff., 146
Discovery, 52ff., 73, 134, 142
Doctrine of Fact, Ch. II passim, Ch. III passim, 63
Dostoievski, 7

Education in Science, 68ff., 110, 130, 133
Einstein, 17n, 53, 71, 85, 87, 102

Emergence, 98ff.
Empiricism, 2, 7, 21ff., 46, 48, 114f.
Epistemology, vii; see also Knowledge, Knowing
Ethics, 11, 18, 106f., 143f., 147ff.
Euclidean Geometry, 58n, 71, 77f., 80f., 93, 118
Evidence, 2, 4, 5, 8n, 9n, 16f., 22, 24, 29ff., 34, 36, 45f., 78f., 92, 100, 104ff.; Ch. VI passim
Existentialism, 9n
Experience, 2ff., 22, 38ff., 45ff., 48ff., 58ff., 103ff., 148ff.
Explanation, 23, 26ff.; see also Causality, Reasons

Facts, 7f., 13, Ch. II passim, 41ff., 46ff., 52ff., 63, 69, 116ff., 131f., 134; see also Truth, Validity
Fallibility of Science, 31f., 45f., 50f., 139f.
Freedom, 39n, 60f., 82, 88
Frege, G., 32f.
Freud, S., 7

Galileo, 10, 12ff., 134ff., 142
Goodman, N., 30n
Greek Science, 9ff., 113
Grünbaum, A., 62n

Habit, 5
Hamlet, 140, 153
Hegel, 39, 40, 41f., 44, 58, 71
Hempel, C., 26n, 30n
Heraclitus, 65
History, 7, 58f., 63ff., 141; of Science, Ch. I passim, 28, 71ff., 135ff., 146f.
Hume, 4, 7f., 15n, 30, 38, 75, 81, 88, 116ff., 130

Ideal of Science, 101ff., 106f., 128ff.
Individuality, 51ff., 69ff., 124, 133ff., 146f.
Induction, 8, 29ff., 75, 115ff., 130

James, W., 4, 34, 89